Materials Development by Additive Manufacturing Techniques

Materials Development by Additive Manufacturing Techniques

Editors

Paolo Fino
Alberta Aversa

MDPI • Basel • Beijing • Wuhan • Barcelona • Belgrade • Manchester • Tokyo • Cluj • Tianjin

Editors
Paolo Fino
Politecnico di Torino
Italy

Alberta Aversa
Politecnico di Torino
Italy

Editorial Office
MDPI
St. Alban-Anlage 66
4052 Basel, Switzerland

This is a reprint of articles from the Special Issue published online in the open access journal *Cells* (ISSN 2073-4409) (available at: https://www.mdpi.com/journal/applsci/special_issues/materials_additive_manufacturing).

For citation purposes, cite each article independently as indicated on the article page online and as indicated below:

LastName, A.A.; LastName, B.B.; LastName, C.C. Article Title. *Journal Name* **Year**, *Volume Number*, Page Range.

ISBN 978-3-03943-032-1 (Hbk)
ISBN 978-3-03943-033-8 (PDF)

Contents

About the Editors

Paolo Fino, Full Professor—Head of Department of Applied Science and Technology (DISAT)—Politecnico di Torino. Prof. Paolo Fino graduated in Chemical Engineering in 1997 at Politecnico di Torino, and completed his PhD in 2001 in Materials Engineering at Politecnico di Torino, in collaboration with Politecnico di Milano. He became an associate professor in 2011 and a full professor in 2014 at Politecnico di Torino, where he teaches Science and Technology of Materials and Materials for Additive Manufacturing (AM). He has been Head of Department of Applied Science and Technology of Politecnico di Torino since 2015. He is the President of CIM4.0: Competence Industry Manufacturing Competence Center of Torino. His main research interests are in the field of additive manufacturing, and in particular of materials for AM processes. He participated, in various roles, in several regional, national and European research projects focused on AM processes and materials, such as TiAl Charger, E-BREAK, Borealis, GETREADY, and many others. He has disparate industrial collaborations with many of the main AM machine producers or end users.

Alberta Aversa, Assistant Professor at Politecnico di Torino. She studied at Università degli Studi di Napoli Federico II, where she completed her bachelor's degree in Materials Science and Engineering in 2010. She then studied at Politecnico di Torino, where she graduated in Materials Engineering in 2013, and completed her PhD in Materials Science and Technology in 2017. She became an assistant professor at Politecnico di Torino in 2018. Her main research interests are materials development for additive manufacturing processes, and she focuses in particular on the characterisation and the developments of new alloys for AM processes. She has participated in several regional and European research projects focused on additive manufacturing processes, such as AMAZE, Borealis, Stamp and MANUELA.

Editorial

Special Issue on Materials Development by Additive Manufacturing Techniques

Alberta Aversa * and Paolo Fino

Department of Applied Science and Technology, Politecnico di Torino, Corso Duca degli Abruzzi 24, 10129 Torino, Italy; paolo.fino@polito.it
* Correspondence: alberta.aversa@polito.it; Tel.: +39-011-090-4763

Received: 21 July 2020; Accepted: 23 July 2020; Published: 25 July 2020

Abstract: Additive manufacturing (AM) processes are steadily gaining attention from many industrial fields, as they are revolutionizing components' designs and production lines. However, the full application of these technologies to industrial manufacturing has to be supported by the study of the AM materials' properties and their correlations with the feedstock and the building conditions. Furthermore, nowadays, only a limited number of materials processable by AM are available on the market. It is, therefore, fundamental to widen the materials' portfolio and to study and develop new materials that can take advantage of these unique building processes. The present special issue collects recent research activities on these topics.

Keywords: additive manufacturing; materials development; mechanical properties; polymers; metals; ceramics

---

1. Introduction

Additive manufacturing (AM) is an innovative class of production technologies, which is often considered to have a large impact in all manufacturing activities, as it allows for the production of complex-shaped components without the need of dedicated tools [1]. This family of production techniques had a large success in recent years, not only thanks to the design freedom that it is possible to achieve, but also thanks to the possibility to produce customized components and to reduce time to market and costs of some production lines [2].

2. Materials for Additive Manufacturing

The countless advantages and challenges of AM processes from a design and from a productivity point of view have been widely discussed in recent years, but, recently, many research studies pointed out that these innovative processing technologies also bring many advantages and challenges from a material perspective [3]. From a materials point of view, in fact, the main issues to be solved are related to the study of the AM parts' properties and to the limited amount of processable materials available on the market.

On the basis of these considerations, many universities, research centers, and industries started studying the correlations between feedstock properties, AM process parameters, and materials properties and are seeking to expand the portfolio of materials available for AM processes [3]. This special issue was, therefore, introduced to summarize the recent research activities on these topics. The main recent advances in AM materials development are described below per each material class.

2.1. Polymers

Polymers are by far the most-used materials in AM due to their simple processes, easy availability and low cost. The most-used polymer AM techniques are stereolithography, selective laser sintering

(SLS), fused deposition modelling (FDM), laminated object manufacturing (LOM), and 3D bioprinting. Each type of technology can process only specific polymers. Photopolymer resins for stereolithography, for example, are the most-used ones in the industrial field, mainly thanks to the excellent accuracy it is possible to achieve through this building process [4]. Polystyrene, polyamide, and thermoplastic elastomers are also widely used and generally processed by SLS. As the mechanical properties of printed polymers seemed to be a major concern, large efforts were carried out to process composites using various AM technologies [5].

Recently, much research was carried out on the development of polymers for the FDM technology. The most common thermoplastic polymers face in fact have many issues, mainly related to their physical properties. In this frame, several studies recently investigated the processability and the properties of Ultem 9085, which is a thermoplastic polymer especially designed by Stratasys for the FDM process [6]. The main advantages of this composition are related to its high glass transition temperature, good flame retardancy, and high mechanical properties. Recent works, reported in this special issue, demonstrated that, due to the layer by layer process, the Ultem tensile properties are strongly anisotropic and heavily related to the building parameters [6,7]. Similarly, Solorio et al. investigated the FDM processability and properties of an innovative amorphous poly(lactide acid) (PLA) blend with poly(styrene-co-methyl methacrylate) (poly(S-co-MMA)) [8]. The study, published in this special issue, demonstrated how the introduction of MMA allowed for an improvement of the processability of the PLA filaments.

2.2. Metals

Nowadays mainly steels, titanium, aluminum, and nickel alloys are successfully processed by AM and used in disparate applications [9]. However, not all alloys belonging to these families can be successfully processed by the most common metal-AM techniques, such as laser powder bed fusion (LPBF), electron beam melting (EBM), and directed energy deposition (DED).

Steel has been by far the first alloy class to be processed and has, therefore, been used in several industries, such as the automotive and aerospace ones [10]. Among AM processable steels, the most studied ones are stainless steel, such as 316L and 304L, precipitation hardening (PH) steels, such as 17-4 PH, tool steels, such as H13 and M2, and maraging steels, such as 18Ni-300. Large efforts have been made in previous works in order to understand the microstructure of steels' built components and their correlation with the building parameters together with the effect that they have on mechanical properties [11]. Saboori et al. summarized the main data and results obtained on DED 316L samples in a review published in the present special issue [11].

Titanium alloys are the most-used alloys in AM thanks to the wide range of applications they have in the biomedical and aerospace fields. Many of these applications can take advantage of the possibility to produce complex and customized parts. Furthermore, the vacuum EBM process of Ti6Al4V Gd23 alloy allows the control of the interstitial content and the consequent respect of the standards.

Aluminum alloys also had large success in the AM field mainly thanks to the strong interest of aerospace companies that need the production of complex, lightweight components [12]. Currently, however, mainly Al–Si alloys, with a near eutectic composition, are processable by AM while most of the Al high-strength alloys strongly suffer from solidification cracking during AM processing. There is, therefore, a limited amount of aluminum alloys processable by AM and, recently, universities, research centers, and companies are investing in the development of new compositions specifically designed for AM, such as the Scalmalloy® or the A20X™ [3].

Nickel alloys, such as In625, In718, and HastelloyX, have been widely used for the AM production of parts that need high-creep and corrosion resistance, such as engine turbine blades, turbochargers, heat exchangers, and petrochemical equipment [13]. The strong interest in this alloy class has pushed the research in the understanding of the microstructure–properties correlation of these materials. Recently other alloys belonging to the Ni family have been successfully processed by AM. As an example, in this special issue, results about the Monel Ni–Cu alloy are reported [14]. This alloy, which has been recently processed by LPBF, showed good processability within specific parameters. High-mechanical properties were measured thanks to fine microstructure and high residual stresses [14].

2.3. Ceramics

Ceramic AM processes are generally classified into direct (or single-step) and indirect (or multistep) methods [15]. In the first class of technologies, the material is fabricated in a single process in which both the final shape and the materials' properties are obtained. These direct methods allow a larger design freedom and are generally, therefore, preferred when complex geometries have to be built. The disadvantage of these processes is that the manufactured parts are usually porous and characterized by a high surface roughness. The processes belonging to the second class, on the contrary, need several steps to reach the final component's consolidation. In the first step, the shape is provided, and the green body is obtained through binding. The subsequent steps are needed to consolidate the part and reach the desired properties. The main advantages of these processes are mainly associated with reduced delamination and anisotropy issues.

The ceramic AM processes have rapidly evolved in recent years, however, in many cases, the mechanical properties of manufactured parts do not reach the desired values [16]. Because of this reason, lately, large efforts have been made to improve the ceramic AM processes' capabilities and to enlarge the palette of processable materials, also involving high-performance ceramics (HPCs) [16,17].

Altun et al., for example, demonstrated, in a paper published in this special issue, the applicability of the indirect AM lithography-based ceramic manufacturing (LCM) method to the production of precise and complex silicon nitride (Si_3N_4) parts. This nonoxide ceramic has attracted large interest thanks to its unique properties, such as high toughness, strength, and thermal shock resistance together with an outstanding biocompatibility that makes it an excellent candidate for dental applications [18].

3. Conclusions

The studies reported in this special issue clearly highlight the importance of the materials' development in AM applications. It is striking that, in most of the cases, a strong correlation between building conditions and materials' properties exist. Furthermore, these studies make it apparent that AM processes open large possibilities in the development of new materials having specific properties and distinct functionalities.

Funding: This research received no external funding.

Acknowledgments: We would like to thank all the authors and peer reviewers for their valuable contributions to this special issue. Thanks are also due to the MDPI management and staff for their editorial support, which increased the success of this special issue.

Conflicts of Interest: The authors declare no conflict of interest.

References

1. DebRoy, T.; Mukherjee, T.; Milewski, J.O.; Elmer, J.W.; Ribic, B.; Blecher, J.J.; Zhang, W. Scientific, technological and economic issues in metal printing and their solutions. *Nat. Mater.* **2019**, *18*, 1026–1032. [CrossRef] [PubMed]
2. Baumers, M.; Dickens, P.; Tuck, C.; Hague, R. The cost of additive manufacturing: Machine productivity, economies of scale and technology-push. *Technol. Forecast. Soc. Chang.* **2016**, *102*, 193–201. [CrossRef]
3. Aversa, A.; Marchese, G.; Saboori, A.; Bassini, E.; Manfredi, D.; Marchese, G.; Saboori, A.; Bassini, E.; Fino, P.; Lombardi, M.; et al. New Aluminum Alloys Specifically Designed for New Aluminum Alloys Specifically Designed for Laser Powder Bed Fusion: A Review. *Materials* **2019**, *12*, 1007. [CrossRef] [PubMed]
4. Ngo, T.D.; Kashani, A.; Imbalzano, G.; Nguyen, K.T.Q.; Hui, D. Additive manufacturing (3D printing): A review of materials, methods, applications and challenges. *Compos. Part B Eng.* **2018**, *143*, 172–196. [CrossRef]
5. Parandoush, P.; Lin, D. A review on additive manufacturing of polymer-fiber composites. *Compos. Struct.* **2017**, *182*, 36–53. [CrossRef]

6. Padovano, E.; Galfione, M.; Concialdi, P.; Lucco, G.; Badini, C. Mechanical and Thermal Behavior of Ultem®9085 Fabricated by Fused-Deposition Modeling. *Appl. Sci.* **2020**, *10*, 3170. [CrossRef]
7. Tosto, C.; Saitta, L.; Pergolizzi, E.; Blanco, I.; Celano, G.; Cicala, G. Methods for the Characterization of Polyetherimide Based Materials Processed by Fused Deposition Modelling. *Appl. Sci.* **2020**, *10*, 3195. [CrossRef]
8. Solorio, A.; Vega, L. Filament Extrusion and Its 3D Printing of Poly (Lactic Acid)/Poly (Styrene-co-Methyl Methacrylate) Blends. *Appl. Sci.* **2019**, *9*, 5153. [CrossRef]
9. Frazier, W.E. Metal Additive Manufacturing: A Review. *J. Mater. Eng. Perform.* **2014**, *23*, 1917–1928. [CrossRef]
10. Bajaj, P.; Hariharan, A.; Kini, A.; Kürnsteiner, P.; Raabe, D.; Jägle, E.A. Steels in additive manufacturing: A review of their microstructure and properties. *Mater. Sci. Eng. A* **2020**, *772*, 138633. [CrossRef]
11. Saboori, A.; Aversa, A.; Marchese, G.; Biamino, S.; Lombardi, M.; Fino, P. Microstructure and Mechanical Properties of AISI 316L Produced by Directed Energy Deposition-Based Additive Manufacturing: A Review. *Appl. Sci.* **2020**, *10*, 3310. [CrossRef]
12. Trevisan, F.; Calignano, F.; Lorusso, M.; Pakkanen, J.; Aversa, A.; Ambrosio, E.; Lombardi, M.; Fino, P.; Manfredi, D. On the Selective Laser Melting (SLM) of the AlSi10Mg Alloy: Process, Microstructure, and Mechanical Properties. *Materials* **2017**, *10*, 76. [CrossRef] [PubMed]
13. Graybill, B.; Li, M.; Malawey, D.; Ma, C.; Alvarado-Orozco, J.M.; Martinez-Franco, E. Additive manufacturing of nickel-based superalloys. In Proceedings of the ASME 2018 13th International Manufacturing Science and Engineering Conference MSEC 2018, College Station, TX, USA, 18–22 June 2018; Volume 1.
14. Raffeis, I.; Adjei-Kyeremeh, F.; Vroomen, U.; Westhoff, E.; Bremen, S.; Hohoi, A.; Bührig-Polaczek, A. Qualification of a Ni–Cu Alloy for the Laser Powder Bed Fusion Process (LPBF): Its Microstructure and Mechanical Properties. *Appl. Sci.* **2020**, *10*, 3401. [CrossRef]
15. Deckers, J.; Vleugels, J.; Kruth, J.P. Additive Manufacturing of Ceramics: A Review. *J. Ceram. Sci. Technol.* **2014**, *5*, 245–260.
16. Wang, J.C.; Dommati, H.; Hsieh, S.J. Review of additive manufacturing methods for high-performance ceramic materials. *Int. J. Adv. Manuf. Technol.* **2019**, *103*, 2627–2647. [CrossRef]
17. Zocca, A.; Colombo, P.; Gomes, C.M.; Günster, J. Additive Manufacturing of Ceramics: Issues, Potentialities, and Opportunities. *J. Am. Ceram. Soc.* **2015**, *98*, 1983–2001. [CrossRef]
18. Altun, A.A.; Prochaska, T.; Konegger, T.; Schwentenwein, M. Dense, strong, and precise silicon nitride-based ceramic parts by lithography-based ceramic manufacturing. *Appl. Sci.* **2020**, *10*, 996. [CrossRef]

Article

Methods for the Characterization of Polyetherimide Based Materials Processed by Fused Deposition Modelling

Claudio Tosto, Lorena Saitta, Eugenio Pergolizzi, Ignazio Blanco, Giovanni Celano and Gianluca Cicala *

Department of Civil Engineering and Architecture, University of Catania, Viale Andrea Doria 6, 95125 Catania, Italy; claudio.tosto@unict.it (C.T.); lorena.saitta@phd.unict.it (L.S.); euper@hotmail.com (E.P.); iblanco@unict.it (I.B.); giovanni.celano@unict.it (G.C.)
* Correspondence: gianluca.cicala@unict.it; Tel.: +39-095-738-2760

Received: 20 April 2020; Accepted: 30 April 2020; Published: 3 May 2020

Featured Application: The present work aims to provide an insight into characterization techniques for Fused Deposition Modelling. The outcomes can guide the development of novel standards for FDM™.

Abstract: Fused deposition modelling (FDM™) is one of the most promising additive manufacturing technologies and its application in industrial practice is increasingly spreading. Among its successful applications, FDM™ is used in structural applications thanks to the mechanical performances guaranteed by the printed parts. Currently, a shared international standard specifically developed for the testing of FDM™ printed parts is not available. To overcome this limit, we have considered three different tests aimed at characterizing the mechanical properties of technological materials: tensile test (ASTM D638), flexural test (ISO 178) and short-beam shear test (ASTM D2344M). Two aerospace qualified ULTEM^{TM} 9085 resins (i.e., tan and black grades) have been used for printing all specimens by means of an industrial printer (Fortus 400mc). The aim of this research was to improve the understanding of the efficiency of different mechanical tests to characterize materials used for FDM™. For each type of test, the influence on the mechanical properties of the specimen's materials and geometry was studied using experimental designs. For each test, 2^2 screening factorial designs were considered and analyzed. The obtained results demonstrated that the use of statistical analysis is recommended to ascertain the real pivotal effects and that specific test standards for FDM™ components are needed to support the development of materials in the additive manufacturing field.

Keywords: polyetherimide; additive manufacturing; fused filament modelling; mechanical properties; design of experiments

1. Introduction

Additive manufacturing (AM) is a layer-by-layer building technique that allows complex shapes to be obtained without the use of a mold. AM is a promising area for manufacturing of components from prototypes to functional structures. The application of AM covers different sectors such as aerospace, automotive, semiconductor and biomedical applications.

Fused filament fabrication (FFF), also known as fused deposition modeling (FDM™), is one of the most popular AM techniques. FDM™ is based on the melting of a thermoplastic filament that is laid on a platform to create each layer on top of the other. The FDM™ process is controlled by many parameters which range from material type to several machine settings such as the nozzle diameter and temperature, printing speed, feed rate, bed temperature, raster angle and width [1].

Several detailed studies are reported in the literature about the influence of the printing settings on the mechanical properties of 3D-printed parts. Es-Said et al. [2] showed that polymer chain alignment occurs during the filament deposition. As a result, the tensile, flexural and impact resistance varies with different raster orientations. Similar results were obtained by Ahn et al. [3]. In their study, the effects of the raster orientation, air gap, bead width, color and model temperature parameters on the tensile strength were evaluated. Results showed that the air gap and raster orientation influence the tensile strength; conversely, the bead width, model temperature and color do not have a significant effect. In another study, Lee et al. [4] concluded that the layer thickness, the raster angle and the air gap influence the elastic performance of 3D-printed ABS (Acrylonitrile Butadiene Styrene) Parts.

The ASTM D638 tensile test and the ASTM D790 or ISO 178 flexural test are both widely used standards for testing polymeric materials processed by injection or compression molding. Thus, practitioners might be interested in extending their implementation to the characterization of the mechanical properties of FDM™ printed parts. Unfortunately, these standards do not account for the presence of voids that are unavoidable in FDM™. In addition, they were not specifically developed to characterize the interlayer bonding which influences the mesostructures of FDM™ printed parts. Tronvoll et al. [5] showed that voids found in FDM™ printed parts significantly impact the tensile properties. According to Sun et al. [6], the chamber temperature and variations in the convection coefficient have a strong effect on the cooling temperature profiles, as well as on the mesostructure and overall quality of the bonding between filaments. However, they did not measure the interlayer strength since the performed flexural tests yielded large variation in the results.

Only a few papers in the AM literature have been focused on the study of the bonding quality between layers and rasters printed by FDM™. Recently, interlaminar bonding has been measured by using the short-beam strength (SBS) test. This test is commonly used for fiber reinforced composites [7–10]. A study of the interlaminar bonding performance of continuous fiber reinforced thermoplastics printed by FDM™ showed a correlation between porosity and the interlaminar shear strength (ILSS) [7]. O' Connor [9] confirmed these findings working with similar materials. In a recent paper, SBS tests indicated improved sensitivity to measure interlaminar bonding effects for different materials compared to tensile or flexural tests [10]. However, all these papers lacked in terms of the statistical analysis of the measured data. Some research tried to rationalize the results of mechanical testing using the design of experiment (DoE) toolbox of statistical techniques [3,11–15]. Vicente et al. [15] showed that the interlayer cooling time can influence the ultimate tensile strength (UTS) because of different bonding properties between the layers. The effect was more pronounced for the shorter Type V sample rather than for the longer Type I sample. However, the effect of the sample type on the interlayer bonding was not systematically discussed by measuring the interlayer bonding. Additionally, tensile testing based on the ASTM D638 has been criticized for dog bone specimens because of the large stress concentrations caused by the termination of the longitudinal roads [3]. ASTM D3039 was proposed to overcome this problem.

In this paper the mechanical properties of two commercial grades of polyetherimides (PEI) are discussed. The paper is organized as follows: first, the two as-received filaments were characterized by thermal analysis to determine differences in the material behavior. Secondly, subsequently printed specimens were analyzed by different mechanical tests ranging from tensile to flexural and SBS. For each material type, the sample dimensions were varied to unveil their effect on the mechanical properties. All results obtained by the tests were statistically analyzed as 2^2 replicated screening designs.

2. Materials and Methods

ULTEM™ 9085, a high temperature thermoplastic blend consisting of PEI and a copolymer to improve the flow, was used in this study. ULTEM™ 9085 is excellent for FDM™ as it shows improved rheology for processing over standard PEI [16]. ULTEM™ 9085 is qualified for aerospace applications. Two ULTEM™ 9085 grades are available from Stratasys classified as tan and black. Additionally, the specifications of the materials differ based on the color itself. The study of the two materials started with thermal characterization. By means of thermal analyses, which are based on the viscoelastic behavior study and the calorimetric glass transition temperature (Tg) determination, we wanted to find out if the two materials show different material properties in general. Based on this finding, in the second step of the investigation a mechanical characterization of the two materials was performed.

To characterize the mechanical behavior of the two ULTEM™ 9085 grades and to understand which mechanical test can be properly used for this kind of FDM™ printed, the combined effect of the material and specimen geometry on the results of different mechanical tests was investigated in our experimental study. To this end, replicated 2^2 screening designs were analyzed for each testing methodology. Two independent variables (factors) were considered in the study: material (factor A) and specimen geometry (factor B). Both factors were varied on two levels. The material is varied at 2 levels by printing either tan or black ULTEM™ 9085. The b = 2 levels for the specimen geometry were selected depending on the test used to get the mechanical properties. For the tensile test (ASTM D638), the b = 2 levels correspond to the Type I and Type IV as defined by the standard. For the flexural (ISO 178) test and the short beam strength (SBS) (ASTM D2344M) test, the b = 2 levels were obtained by printing bars with different lengths (L) (i.e., L_1 = 122 mm and L_2 = 165 mm). The choice of these two values for L was motivated by achieving a right trade-off between the specimen length required by the tensile test and the specimen length fixed by the flexural or SBS test. The reason why we decided to investigate the effect of the specimen geometry was to consider the effect of interlayer cooling. In fact, as reported by the literature, the weld temperature decreases at a rate of approximately 100 °C/s and it remains above the glass transition temperature for about 1 s [17]. As a consequence of this cooling process, printing samples with different lengths can lead to a different temperature profile within the printed parts and, therefore, to a different interlayer bonding strength. This phenomenon is shown in [18], where the part length significantly influences the warpage due to thermal induced stresses. Once the factors (independent variables) were identified in the experimental plan, the dependent variables (responses) were selected. For the tensile test, we considered the UTS and the Young's modulus as the responses to be investigated. Similarly, we took the flexural stress and the ILSS as responses for the flexural and SBS test, respectively. For each experimental study, the number of replications were set equal to n = 5. Therefore, N = a · b · n = 20 runs were carried out for each experimental plan. The statistical analysis of the experimental plan was performed by using the commercial Design-Expert software (Stat-Ease, Minneapolis, US). Table 1 summarizes the information about the three experimental plans.

Table 1. Experimental plans. Factors, levels and responses for each investigated test.

Test	Standard	Factor	Symbol	Type	Unit	Low Level (−1)	High Level (+1)
Tensile	ASTM D638	Material	A	Categorical	-	ULTEM™ 9085 Tan	ULTEM™ 9085 Black
		Geometry	B	Categorical	-	Type I	Type IV
Flexural	ISO 178	Material	A	Categorical	-	ULTEM™ 9085 Tan	ULTEM™ 9085 Black
		Geometry	B	Categorical	mm	122	165
SBS	ASTM D2344M	Material	A	Categorical	-	ULTEM™ 9085 Tan	ULTEM™ 9085 Black
		Geometry	B	Categorical	mm	122	165

The specimens were printed on the FDM™ machine trademarked as Fortus 400mc (Stratasys, Los Angeles, CA, USA). The printing volume is (406 · 356 · 406) mm³. The chamber is heated when printing engineering polymers such as PEI to minimize the thermal distortion.

The specimen's geometry was printed according to the different mechanical testing standards used throughout the manuscript (Figure 1).

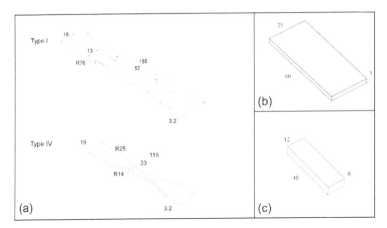

Figure 1. Dimensions (in mm) of the specimens. (**a**) Tensile test specimens (ASTM D638 type I, IV); (**b**) flexural test specimens; (**c**) short-beam shear specimens.

The selected printing settings are summarized in Table 2. These parameters were selected according to past experience to minimize the presence of internal voids [19]. All the specimens were oriented flatwise on the XY plane. To avoid negative notch effects leading to premature failure, as reported in some previous research [3], the start and stop positions for printing the tensile specimens were set in one corner in the grip zone (Figure 2).

Table 2. Printing conditions for the preparation of the specimens.

Parameters	Unit	Value
Infill	%	100
Infill type		Solid
Support type		ULTEM Support
Raster angle	deg	0/90
Layer height	μm	254
Tip		T16
Shrink factor (x)		1.01
Shrink factor (y)		1.01
Shrink factor (z)		1.0097
Contours width	mm	0.508
Part raster width	mm	0.508
Raster to raster air gap		0
Contour to raster air gap		0
Contour to contour air gap		0

The viscoelastic behavior of the two material types was investigated using a DMA Tritec 2000 (Triton Technology Ltd., Nottinghamshire, UK) by single cantilever geometry and sample size $(10 \cdot 5 \cdot 2)$ mm^3. The tests were carried out at 1 and 10 Hz with 2 °C/min heating rate from 25 °C to 250 °C.

A Shimadzu DSC 60 (Shimadzu, Kyoto, Japan) was used for calorimetric glass transition temperature (Tg) determinations. The apparatus was calibrated in enthalpy and temperature by following the procedure discussed in [20]. Afterwards, the enthalpy calibration was checked by the melting of fresh indium, showing an agreement with the literature standard within 0.25% [21]. This happened while the temperature calibration was checked by several scans with fresh indium and tin, showing an agreement within 0.08% with respect to the literature values [21]. The DSC scans have been performed on samples of about $6.0 \cdot 10^{-3}$ g, held in sealed aluminum crucibles at a heating rate of 10 °C/min and static air atmosphere. The investigations were carried out in a range of temperatures from room temperature up to 300 °C and each scan was performed in triplicate. The considered values were averaged from those of three runs, the maximum difference between the average and the experimental values being within ±1 °C.

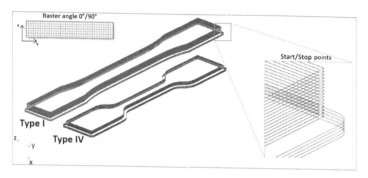

Figure 2. Slice and toolpath for tensile test specimens.

The mechanical properties of printed specimens were measured by using an Instron 5985 universal testing machine (Instron, Milan, Italy) equipped with a load cell of 10 kN. For each test, the tools required for the various standard tests were installed. System control and data collection were performed using the Blue Hill 3.61 software (Instron, MA, USA). Following the DoE method, we randomized the testing order for all samples and test types.

Tensile specimens were tested according to ASTM D638. The test was carried out in strain control mode at a speed of 2 mm/min, using a clip extensometer with 25 mm useful length. Tensile specimens were printed with Type I and Type IV geometry, as specified in the ASTM D638 standard (Figure 1a).

The flexural test (ISO 178) was performed with $(60 \cdot 25 \cdot 3)$ mm^3 samples (Figure 1b) and a span length (distance between supports) equal to 48 mm. The tests were conducted at a speed of 2 mm/min. The flexural samples were obtained by cutting bars with length equal to 122 mm and 165 mm in pieces having a standard length of 60 mm.

For the ILSS (ASTM D2344M), samples of size $(40 \cdot 12 \cdot 6)$ mm^3 were considered, with a span length of 24 mm (Figure 1c). ILSS tests were carried out at a speed of 1 mm/min. The ILSS samples were obtained by cutting bars having length equal to 122 mm and 165 mm in pieces with a standard length of 40 mm.

Scanning electron microscopy micrographs were obtained with a SEM EVO-MA15 by Zeiss, Cambridge (UK). The fractured surfaces were sputter coated with gold before the SEM micrograph was taken.

3. Results and Discussion

3.1. Thermal Characterization

A preliminary study on tan and black ULTEM™ 9085 materials was carried out to define the difference in terms of viscoelastic and thermocalorimetric behavior. Previous tests on ULTEM™ 9085 have shown that it is a PEI modified polymer containing a copolymer for improved flow [16]. The tan versus temperature plot is reported in Figure 3 for both polymers. A wide peak at 185 °C and a shoulder at 140 °C were observed for the tan sample. For the black sample, the peak and the shoulder shifted to 195 °C and 148 °C, respectively.

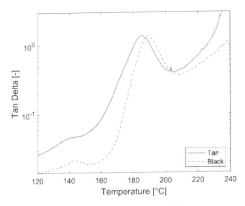

Figure 3. Tanδ versus temperature for ULTEM™ 9085 tan and black.

DSC data showed similar results for tan and black materials, with a glass transition observed at around 180 °C (Figure 4). The tan sample showed two distinct thermal transitions while only one was observed for the black resin. Similar results for PEI blends were observed in the past [22]. However, the DSC test seems unable to clearly resolve the thermal transitions as observed in the DMA test, despite the fact that the behavior is also different for the two grades for this analysis.

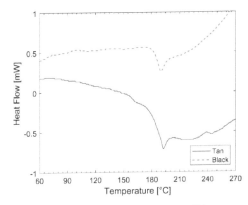

Figure 4. Differential scanning calorimetry for ULTEM™ 9085 tan and black.

The thermal analyses reported here show that the two materials have a different behavior despite being quite similar in composition. Filament pigmentation was reported to impact on the finish and the mechanical behavior of PLA based filaments [23–26]. However, similar data were not reported previously for PEI based filaments. Therefore, the study was continued by characterizing the mechanical behavior of the printed parts with the two materials.

3.2. Mechanical Characterization

The mechanical characterization of the investigated materials requires the implementation of a proper test. Unfortunately, an accepted international standard specifically developed for the testing of the mechanical properties of FDM™ printed parts is not available yet. For this reason, we considered and compared the performance of three well-known mechanical tests available in the literature for other fields of application. The objective was finding a proper test for characterizing the two 3D-printed ULTEM™ XY material types by analyzing different experimental plans.

3.2.1. Tensile Testing

After generating the experimental plan and collecting the response observations (Table S1) of the tensile test according to the ASTM D638 standard (UTS and Young's modulus), an ANOVA study was performed using the Design-Expert software. Randomization was used for the testing sequence, as reported in Table S1 in the Supplementary Material. The average tensile stress of the five tested samples versus displacement curves are shown in Figure 5. All the tested specimens showed brittle failure with no yielding. The UTS varied in the range between 48.99 MPa and 61.98 MPa for the two materials. Young's modulus varied in the range between 2.05 GPa and 2.34 GPa. The measured tensile properties were similar to those reported in other papers focusing on ULTEM™ 9085 [16,27–29]. Zaldivar et al. [29] showed tensile strength values varying from 46.83 MPa for flat samples to 71.03 MPa for on-edge samples. The tensile modulus varied from 1.77 GPa to 2.48 GPa. In this study, the raster orientation varied from 90° to 0°. Similarly, Byberg et al. [28] reported tensile strength values from 31.30 MPa to 70.60 MPa. FDM™ samples show lower mechanical properties, in particular the UTS reduction ranges between 20–40% and the strain of about 2% [30]. These findings depend on the presence of voids and on the thermal history of the printed samples when compared to injection molded specimens.

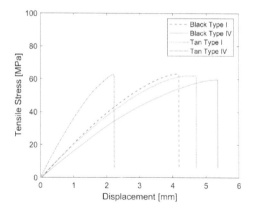

Figure 5. Average tensile stress versus displacement curves.

The Analysis of Variance (ANOVA) table for Young's modulus response is shown in Table 3. Model adequacy checking on the residuals from the analysis did not show any anomaly, as shown in Figure 6 The material type (factor A) is an influential factor (p-value < 0.001) on Young's modulus (Figure 7, Table 3) and it is involved in a significant interaction AB (p-value < 0.001) with the geometry (factor B) (Figure 8). The model appears to have a good robustness to define the observed response with a high R-squared value of 0.8368. Conversely, when the UTS response is considered as the response variable, the ANOVA analysis shows that the material and the geometry factors do not influence it (Table 4). The tensile test is actually unable to characterize the ultimate tensile strength (UTS) for the two materials, as revealed by the very low R-squared value of 0.0764 We explain this finding by considering that the tensile test for flatwise printed specimens is not as sensitive to the interlayer bonding as it is for the upright orientation case where interlayers are directly loaded. In fact, for flatwise samples the longitudinally oriented rasters can sustain applied loads.

ULTEM™ 9085 displayed a structure with a clear distinction of the deposited filaments that are not completely bonded and melted together (Figure 9). Therefore, testing methods that account for the interlayer bonding resistance should be used to fully characterize the mechanical behavior of the material. The morphological analysis of the fractured specimen reveals other features. The longitudinal rasters that were aligned along the tensile load show a deformed cross section with some yielding before failure (see green arrows), and the transverse rasters were not deformed and there were some zones where adhesive failures occurred (see red ellipses). It is important to notice that the yielding occurs locally and for the longitudinal raster only. This is not reflected in the macroscopic behavior of the samples as shown in Figure 6. Compared to other studies, the level of fiber-to-fiber fusion seems lower for the analyzed specimen [29]. Crack propagation seems to also depend on the raster orientation [31]. This analysis highlights the importance of characterizing the interlayer bonding for these samples.

Table 3. ANOVA table for tensile test (response is Young's modulus).

Source	Sum of Squares	df	Mean Square	F Value	*p*-Value
Model	0.1109	3	0.0370	25.6417	<0.0001
A–Material	0.0575	1	0.0575	39.8659	<0.0001
B–Geometry	0.0056	1	0.0056	3.8861	0.0674
AB	0.0575	1	0.0575	39.8659	<0.0001
Pure Error	0.0216	15	0.0014		
Cor. Total	0.1325	18			
Std. Dev.	0.038	**R-squared**	0.8368		
Mean	2.19	**Adj. R-squared**	0.8042		
C.V. %	1.73	**Pred. R-squared**	0.7442		

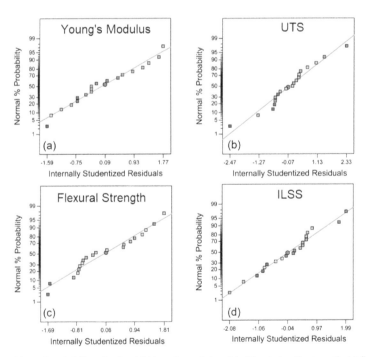

Figure 6. Normal probability plot for: (**a**) Young's modulus; (**b**) ultimate tensile strength; (**c**) flexural strength; and (**d**) interlaminar shear strength.

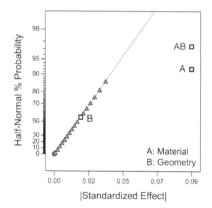

Figure 7. Normal probability plot for tensile test (Young's modulus).

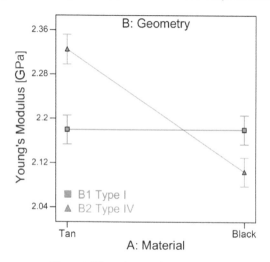

Figure 8. Effects diagram for tensile test.

Table 4. ANOVA table for tensile test (response: UTS).

Source	Sum of Squares	df	Mean Square	F Value	*p*-Value
Model	12.4925	3	4.1642	0.4413	0.7266
A–Material	1.7387	1	1.7387	0.1843	0.6735
B–Geometry	2.0563	1	2.0563	0.2179	0.6469
AB	8.6975	1	8.6975	0.9218	0.3513
Pure Error	150.9726	16	9.4358		
Cor. Total	163.4652	19			
Std. Dev.	3.0718	**R-squared**	0.0764		
Mean	59.1469	**Adj. R-squared**	−0.0967		
C.V. %	5.1935	**Pred. R-squared**	−0.4430		

Figure 9. Fracture surface morphology for a tensile sample (tan resin). Red ellipse highlights the adhesive failure on the transverse rasters. The green arrow highlights the yielded longitudinal rasters.

3.2.2. Flexural Testing

The ISO 178 flexural test is considered as a testing method allowing some of the limitations of tensile testing to be overcome because of the absence of severe constraints due to the clamping of the samples [32]. The flexural test investigated in this study was not applied to a tubular geometry as in Kuznetson et al. [32], but it was performed according to the standard ISO 178. The reason for this choice is that the tubular geometry limits the possibility of varying raster orientation in the printed sample. Therefore, the standard ISO 178 geometry was used, as reported in Figure 1, allowing us to use the same raster orientation as for the samples subjected to tensile load. The experimental curves obtained from the flexural test do not show any significant differences between specimens obtained from bars printed with different lengths (Figure 10). The readings of maximum flexural stress varied in the range between 77.48 MPa and 108.02 MPa for the two materials (Table S2). Gebisa et al. [27] in their study reported flexural stresses varying from 52.89 MPa to 126.30 MPa. Although the material type seems to be the only relevant factor (p-value = 0.0379) in this experimental study (Table 5), the small R-squared = 0.22 and adjusted R-squared = 0.17 values obtained from the ANOVA analysis shows that only a small fraction of total variability measured in the flexural stress is due to the investigated factors. Similarly to the tensile test, this reveals a high level of noise affecting the flexural stress which dramatically affects the test sensitivity when applied to FDM™ printed specimens. This result can be explained by the fact that the shear stresses developing within the specimen during a flexural test can influence its results. For this reason, its effect is minimized in the ISO 178 standard by fixing the ratio (r) of support length (L_S) to the specimen height (h) to be equal to 16 [33]. Clearly, this condition is not favorable for the purpose of the mechanical characterization of FDM™ samples where the interlaminar bonding—whose resistance can be tested by the presence of shear stress—plays a relevant role on the mechanical properties of the specimens.

3.2.3. Short-Beam Shear Testing

Among the different test options typically used to characterize fiber-reinforced polymers, the SBS test is a valid option to easily determine the ILSS. For this test, the span-to-thickness ratio is fixed at values in the order that the occurring shear stresses within the specimen are high compared to the normal stresses generated by the bending moment [33]. The average ILSS versus displacement curves obtained for the ULTEM™ 9085 samples are shown in Figure 11 (also see Table S3). From these curves, it is immediately evident the effect of the material type, with the black resin showing higher SBS than the tan resin. For all the tested specimens, the readings of SBS varied in the range between 11.82 MPa and 16.74 MPa. The results of the ANOVA analysis for the experimental plan and the normal probability plot of the effects are shown in Table 6 and Figure 12, respectively. Model adequacy checking on the residuals from the analysis did not show any anomaly. As expected, the material type is clearly the influent factor (p-value < 0.0001) on the ILSS. Neither the geometry nor the second-order interaction between the material type and geometry were significant. However, a very high portion of variability (more than 90%), is found for the material type with R-squared = 0.92 and adjusted R-squared = 0.92. The main effects diagram shown in Figure 13 on the material type factor clearly shows its influence. The two black square points and the intervals on the diagram correspond to the average of the ILSS observations and the 95% confidence intervals for the mean ILSS for tan and black, respectively. The same result was obtained when plotting the main effects diagram for level 165 mm (not shown here).

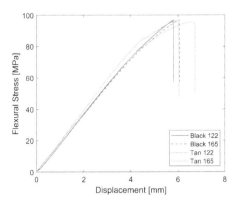

Figure 10. Average flexural stress versus displacement curves.

Table 5. ANOVA table for flexural test (response: flexural strength).

Source	Sum of Squares	df	Mean Square	F Value	*p*-Value
Model	385.27	1	385.27	5.02	0.379
A–Material	385.27	1	385.27	5.02	0.379
Residual	1381.04	18	76.72		
Lack of Fit	72.47	2	36.24	0.44	0.6497
Pure Error	1308.56	16	81.79		
Cor. Total	1766.30	19			
Std. Dev.	8.76	**R-squared**	0.2181		
Mean	95.89	**Adj. R-squared**	0.1747		
C.V. %	9.13	**Pred. R-squared**	0.0347		

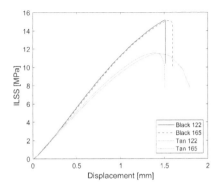

Figure 11. Average ILSS versus displacement curves.

Table 6. ANOVA table for ILSS test (response: short-beam strength).

Source	Sum of Squares	df	Mean Square	F Value	*p*-Value
Model	57.83	1	57.83	217.84	<0.0001
A–Material	57.83	1	57.83	217.84	<0.0001
Residual	4.78	18	0.27		
Lack of Fit	0.75	2	0.37	1.48	0.2569
Pure Error	4.03	16	0.25		
Cor. Total	62.61	19			
Std. Dev.	0.52	R-squared	0.9237		
Mean	14.07	Adj. R-squared	0.9194		
C.V. %	3.66	Pred. R-squared	0.9058		

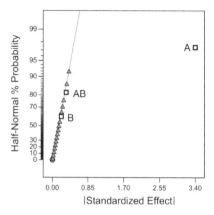

Figure 12. Normal probability plot for SBS test.

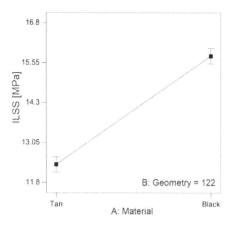

Figure 13. Effects diagram for SBS test.

4. Conclusions

This paper focused on the characterization of two grades of commercial PEI-based filaments used for FDM™, which are tan and black aerospace qualified ULTEM™ 9085. The study of the two materials included both their thermal and mechanical characterization. From the thermal analyses (dynamic mechanical analysis and differential scanning calorimetry) we found that the two material types show a significantly different temperature-dependent behavior. Regarding the mechanical characterization, the absence of a proper test for FDM™ printed specimens led us to the comparison of three different tests: tensile, flexural and short-beam shear. Design of experiment techniques were used to perform the experimental study. An industrial machine (Fortus 400mc) was used for printing high quality specimens. Among the three investigated tests, only the short-beam shear test was able to sufficiently discriminate between the material types. This result strictly depends on the test configuration that privileges the effect of the shear stress internal to the specimen under the loading condition, and the key role played by the interlaminar bonding in the mechanical properties of FDM™ printed parts.

More research is needed to address the correlation between printing parameters and the mechanical properties of printed materials. The need for improving the understanding of correlations and for enlightening the anisotropic behavior is of utmost importance in view of the increased use of reinforced materials in FDM™ to satisfy the need for structural components. Mechanical tests such as double cantilever beam (DCB) or end-notched flexure (ENF) test could also be considered in future research in view of designing a new test standard for FDM™. Additional improved tensile testing with geometries specifically designed to account for material's orthotropy and FDM™ building procedures should be developed. Tapped tensile specimens normally used for fiber reinforced samples could be a solution worth investigating. In terms of future applications, properly developed testing methods would allow for data sets that are useful for easy design of available parts. A standardized test is also needed to have robust techniques for the validation of materials for FDM™ under development.

Supplementary Materials: The following are available online at http://www.mdpi.com/2076-3417/10/9/3195/s1: Table S1, Table S2 and Table S3.

Author Contributions: Conceptualization, G.C. (Giovanni Celano) and G.C. (Gianluca Cicala); methodology, G.C. (Giovanni Celano); software, C.T.; investigation, C.T, L.S., I.B. and E.P.; writing—original draft preparation, G.C. (Giovanni Celano), G.C. (Gianluca Cicala), C.T.; writing—review and editing, G.C. (Giovanni Celano) and C.T.; supervision, G.C. (Gianluca Cicala); project administration, G.C. (Gianluca Cicala). All authors have read and agreed to the published version of the manuscript.

Funding: This research was funded by MIUR, grant number 20179SWLKA Project Title Multiple Advanced Materials Manufactured by Additive technologies (MAMMA) under the PRIN funding Scheme, Project Coordinator G.C.2 and under the funding scheme Change supported by the University of Catania Project Coordinator G.C.2. Claudio Tosto acknowledge the funding of his PhD by MIUR within the PON Ricerca e Innovazione 2014–2020 Asse I "Investimenti in Capitale Umano" -Azione I.1 "Dottorati Innovativi Con Caratterizzazione Industriale" Project Title "Advanced Materials by Additive manufacturing" (AMA).

Conflicts of Interest: The authors declare no conflict of interest. The funders had no role in the design of the study; in the collection, analyses, or interpretation of data; in the writing of the manuscript, or in the decision to publish the results.

Abbreviations

Additive manufacturing	(AM)
American Society for Testing and Materials	(ASTM)
Analysis of variance	(ANOVA)
Design of experiments	(DoE)
Fused deposition modeling	(FDM™)
Fused filament fabrication	(FFF)
Interlaminar shear strength	(ILSS)
International Organization for Standardization	(ISO)
Polyetherimide	(PEI)
Short-beam strength	(SBS)
Ultimate tensile strength	(UTS)

References

1. Harris, M.; Potgieter, J.; Archer, R.; Arif, K.M. Effect of material and process specific factors on the strength of printed parts in fused filament fabrication: A review of recent developments. *Materials* **2019**, *12*, 1664. [CrossRef] [PubMed]
2. Es-Said, O.S.; Foyos, J.; Noorani, R.; Mendelson, M.; Marloth, R.; Pregger, B.A. Effect of Layer orientation on mechanical properties of rapid prototyped samples. *Mater. Manuf. Process.* **2000**, *15*, 107–122. [CrossRef]
3. Ahn, S.H.; Montero, M.; Odell, D.; Roundy, S.; Wright, P.K. Anisotropic material properties of fused deposition modeling ABS. *Rapid Prototyp. J.* **2002**, *8*, 248–257. [CrossRef]
4. Lee, B.H.; Abdullah, J.; Khan, Z.A. Optimization of rapid prototyping parameters for production of flexible ABS object. *J. Mater. Process. Technol.* **2005**, *169*, 54–61. [CrossRef]
5. Tronvoll, S.A.; Welo, T.; Elverum, C.W. The effects of voids on structural properties of fused deposition modelled parts: A probabilistic approach. *Int. J. Adv. Manuf. Technol.* **2018**, *97*, 3607–3618. [CrossRef]
6. Lin, B.; Sundararaj, U. Visualization of polyetherimide and polycarbonate blending in an internal mixer. *J. Appl. Polym. Sci.* **2004**, *92*, 1165–1175. [CrossRef]
7. Caminero, M.A.; Chacón, J.M.; García-Moreno, J.M.R.I. Interlaminar bonding performance of 3D printed continuous fibre reinforced thermoplastic using FDM. *Polym. Test.* **2018**, *68*, 415–423. [CrossRef]
8. Zhang, W.; Cotton, C.; Sun, J.; Heider, D.; Gu, B.; Sun, B.; Chou, T.W. Interfacial bonding strength of short carbon fiber/acrylonitrile-butadiene-styrene composites fabricated by fused deposition modeling. *Compos. Part B Eng.* **2018**, *137*, 51–59. [CrossRef]
9. O'Connor, H.J.; Dowling, D.P. Low-pressure additive manufacturing of continuous fiber-reinforced polymer composites. *Polym. Compos.* **2019**, *40*, 4329–4339. [CrossRef]
10. Caminero, M.Á.; Chacón, J.M.; García-Plaza, E.; Núñez, P.J.; Reverte, J.M.; Becar, J.P. Additive manufacturing of PLA-based composites using fused filament fabrication: Effect of graphene nanoplatelet reinforcement on mechanical properties, dimensional accuracy and texture. *Polymers* **2019**, *11*, 799. [CrossRef]
11. Mercado-Colmenero, J.M.; Rubio-Paramio, M.A.; la Rubia-Garcia, M.D.; Lozano-Arjona, D.; Martin-Doñate, C. A numerical and experimental study of the compression uniaxial properties of PLA manufactured with FDM™™ technology based on product specifications. *Int. J. Adv. Manuf. Technol.* **2019**, *103*, 1893–1909. [CrossRef]
12. Zaman, U.K.; Boesch, E.; Siadat, A.; Rivette, M.; Baqai, A.A. Impact of fused deposition modeling (FDM) process parameters on strength of built parts using Taguchi's design of experiments. *Int. J. Adv. Manuf. Technol.* **2019**, *101*, 1215–1226. [CrossRef]
13. Motaparti, K.P.; Taylor, G.; Leu, M.C.; Chandrashekhara, K.; Castle, J.; Matlack, M. Experimental investigation of effects of build parameters on flexural properties in fused deposition modelling parts. *Virtual Phys. Prototyp.* **2017**, *12*, 1–14. [CrossRef]
14. Jiang, S.; Liao, G.; Xu, D.; Liu, F.; Li, W.; Cheng, Y.; Li, Z.; Xu, G. Mechanical properties analysis of polyetherimide parts fabricated by fused deposition modeling. *High Perform. Polym.* **2019**, *31*, 97–106. [CrossRef]
15. Vicente, C.M.S.; Martins, T.S.; Leite, M.; Ribeiro, A.; Reis, L. Influence of fused deposition modeling parameters on the mechanical properties of ABS parts. *Polym. Adv. Technol.* **2020**, *31*, 501–507. [CrossRef]
16. Cicala, G.; Ognibene, G.; Portuesi, S.; Blanco, I.; Rapisarda, M.; Pergolizzi, E.; Recca, G. Comparison of Ultem 9085 used in fused deposition modelling (FDM) with polytherimide blends. *Materials* **2018**, *11*, 258. [CrossRef] [PubMed]
17. Seppala, J.E.; Migler, K.D. Infrared thermography of welding zones produced by polymer extrusion additive manufacturing. *Addit. Manuf.* **2016**, *12*, 71–76. [CrossRef]
18. Armillotta, A.; Bellotti, M.; Cavallaro, M. Warpage of FDM™™ parts: Experimental tests and analytic model. *Robot Comput. Integr. Manuf.* **2018**, *50*, 140–152. [CrossRef]
19. Cicala, G.; Latteri, A.; Del Curto, B.; Lo Russo, A.; Recca, G.; Farè, S. Engineering thermoplastics for additive manufacturing: A critical perspective with experimental evidence to support functional applications. *J. Appl. Biomater. Funct. Mater.* **2017**, *15*, 10–18. [CrossRef]
20. Blanco, I.; Oliveri, L.; Cicala, G.; Recca, A. Effects of novel reactive toughening agent on thermal stability of epoxy resin. *J. Therm. Anal. Calorim.* **2012**, *108*, 685–693. [CrossRef]

21. Della Gatta, G.; Richardson, M.J.; Sarge, S.M.; Stølen, S. Standards, calibration, and guidelines in microcalorimetry. Part 2. Calibration standards for differential scanning calorimetry* (IUPAC Technical Report). *Pure Appl. Chem.* **2006**, *78*, 1455–1476. [CrossRef]

22. Blanco, I.; Cicala, G.; Ognibene, G.; Rapisarda, M.; Recca, A. Thermal properties of polyetherimide/polycarbonate blends for advanced applications. *Polym. Degrad. Stab.* **2018**, *154*, 234–238. [CrossRef]

23. Valerga, A.P.; Batista, M.; Salguero, J.; Girot, F. Influence of PLA filament conditions on characteristics of FDM parts. *Materials* **2018**, *11*, 1322. [CrossRef] [PubMed]

24. Breda, S.J.; João, F.; Pinto, S.F.; Liane, R.; Henrique, C.L. Analysis of the influence of polylactic acid (PLA) colour on FDM™™ 3D printing temperature and part finishing. *Rapid Prototyp. J.* **2018**, *24*, 1305–1316. [CrossRef]

25. Wittbrodt, B.; Pearce, J.M. The effects of PLA color on material properties of 3-D printed components. *Addit. Manuf.* **2015**, *8*, 110–116. [CrossRef]

26. Cicala, G.; Giordano, D.; Tosto, C.; Filippone, G.; Recca, A.; Blanco, I. Polylactide (PLA) filaments a biobased solution for additive manufacturing: Correlating rheology and thermomechanical properties with printing quality. *Materials* **2018**, *11*, 1191. [CrossRef] [PubMed]

27. Gebisa, A.W.; Lemu, H.G. Investigating effects of Fused-deposition modeling (FDM) processing parameters on flexural properties of ULTEM 9085 using designed experiment. *Materials* **2018**, *11*, 500. [CrossRef]

28. Byberg, K.I.; Gebisa, A.W.; Lemu, H.G. Mechanical properties of ULTEM 9085 material processed by fused deposition modeling. *Polym. Test.* **2018**, *72*, 335–347. [CrossRef]

29. Zaldivar, R.J.; Witkin, D.B.; McLouth, T.; Patel, D.N.; Schmitt, K.; Nokes, J.P. Influence of processing and orientation print effects on the mechanical and thermal behavior of 3D-Printed ULTEM^{TM} 9085 Material. *Addit. Manuf.* **2017**, *13*, 71–80. [CrossRef]

30. El-Gizawy, A.S.; Corl, S.; Graybill, B. Process-induced properties of FDM™™ products. In Proceedings of the 2011 International Conference on Mechanical Engineering and Technology (ICMET 2011), London, UK, 24–25 November 2011.

31. Torrado, A.R.; Roberson, D.A. Failure analysis and anisotropy evaluation of 3D-printed tensile test specimens of different geometries and print raster patterns. *J. Fail. Anal. Prev.* **2016**, *16*, 154–164. [CrossRef]

32. Kuznetsov, V.E.; Solonin, A.N.; Urzhumtsev, O.D.; Schilling, R.; Tavitov, A.G. Strength of PLA components fabricated with fused deposition technology using a desktop 3D printer as a function of geometrical parameters of the process. *Polymers* **2018**, *10*, 313. [CrossRef] [PubMed]

33. Grellmann, W.; Seidler, S. *Polymer Testing*; Springer: Berlin/Heidelberg, Germany, 2013.

Article

Mechanical and Thermal Behavior of Ultem® 9085 Fabricated by Fused-Deposition Modeling

Elisa Padovano [1,*], Marco Galfione [1], Paolo Concialdi [2], Gianni Lucco [2] and Claudio Badini [1]

[1] Politecnico di Torino, Department of Applied Science and Technology, Corso Duca degli Abruzzi 24, 10129 Torino, Italy; s253331@studenti.polito.it (M.G.); claudio.badini@polito.it (C.B.)

[2] Pininfarina S.p.A., Via Nazionale 30, 10020 Cambiano, Italy; p.concialdi@pininfarina.it (P.C.); g.lucco@pininfarina.it (G.L.)

* Correspondence: elisa.padovano@polito.it; Tel.: +39-0110904708

Received: 30 March 2020; Accepted: 28 April 2020; Published: 1 May 2020

Featured Application: Ultem 9085 is a relative new material with well-known flame-retardant properties that has many applications in digital manufacturing and rapid prototyping. Thanks to its high mechanical performance, this material has potential applications in many fields, especially aerospace, automotive, and military industries which require a high strength-to-weight ratio.

Abstract: Fused-deposition modeling (FDM) is an additive manufacturing technique which is widely used for the fabrication of polymeric end-use products in addition to the development of prototypes. Nowadays, there is an increasing interest in the scientific and industrial communities for new materials showing high performance, which can be used in a wide range of applications. Ultem 9085 is a thermoplastic material that can be processed by FDM; it recently emerged thanks to such good properties as excellent flame retardancy, low smoke generation, and good mechanical performance. A deep knowledge of this material is therefore necessary to confirm its potential use in different fields. The aim of this paper is the investigation of the mechanical and thermal properties of Ultem 9085. Tensile strength and three-point flexural tests were performed on samples with XY, XZ, and ZX building orientations. Moreover, the influence of different ageing treatments performed by varying the maximum reached temperature and relative humidity on the mechanical behavior of Ultem 9085 was then investigated. The thermal and thermo-oxidative behavior of this material was also determined through thermal-gravimetric analyses.

Keywords: fused-deposition modeling; mechanical properties; thermal behavior

---

1. Introduction

Additive manufacturing (AM) refers to an innovative technology used to fabricate three-dimensional components starting from a computer-aided design(CAD) model; this terminology was introduced by Charles Hull in 1986 and it was originally devoted to the production of prototypes. Recently, AM has attracted the growing interest of both scientific and industrial communities. In fact, thanks to its flexibility and ease of use, AM has progressively found its way into many manufacturing industries; in addition, it is one of the main topics of many researchers' studies. This innovative technology uses a conceptual approach which is completely different to that adopted by subtractive manufacturing methods, which start from a block of material and progressively remove part of this material through cutting, drilling, and grinding. These latter are typically used to create components for prototyping, manufacturing tooling, and end-use parts which require tight tolerances or geometries which are difficult to produce by molding, casting, or other traditional manufacturing techniques.

On the contrary, AM is based on an additive principle, which allows objects to be made starting from 3D-model data and using a layer-by-layer strategy to build up the desired part. There are many are the advantages connected with this technology, such as the possibility of manufacturing complex geometries, product customization and minimization of waste and scraps, in addition to a reduction in production costs as well as a simplification of the manufacturing cycle [1–3].

Fused-deposition modeling (FDM) was one of the first AM techniques introduced into the market for the production of polymeric components. It is defined as "a material extrusion process used to make thermoplastic parts through a heated extrusion and deposition of materials layer by layer" [4]. A continuous filament is heated at the nozzle, extruded, and then deposited to form stacked layers to build up the final component. The simplicity of the process and its low cost associated to its high speed are the main advantages of FDM. The main limitations of this technique are the anisotropic behavior of printed parts and their poor accuracy and surface finishing [5–7]; moreover, the formation of voids during the filament deposition is frequently observed [8–10]. This results in an increased porosity of final components and in a worsening of their mechanical properties. However, many studies have been performed in order to overcome these limitations.

Various materials are currently processed through FDM, such as acrylonitrile butadiene styrene (ABS), polylactide acid (PLA), polycarbonate (PC), polyamide (PA), polyphenylene sulphide (PPS); polyetherimide (PEI), and polyether ether ketone (PEEK) [11,12]. However, the development of new materials with customized properties represents a key challenge to further extend the potential large-scale application of FDM technology.

Ultem 9085 is a high-performance thermoplastic polymer manufactured by Stratasys which can be processed by FDM; it is a mixture of a polyetherimide (PEI) with a polycarbonate (PC) copolymer blend; the latter is added in order to improve the material flow [13].

It shows high glass transition temperature, excellent flame retardancy, low smoke generation, and good mechanical properties [14]. This set of properties makes this material a promising candidate in different application fields, especially those which require low mass and high strength such as aerospace, marine and automotive sectors. As Ultem 9085 is a relative recent material, no extensive literature is available. There are however, two main aspects that are more deeply investigated in the literature due to their large impact on the mechanical performance of printed parts. Firstly, the quality of final components is greatly influenced by the used-process parameters. FDM involves a quite complex mechanism based on the interaction of different parameters such air gap, raster width, raster angle, contour number, and contour width. Their optimization and the study of how these parameters can individually or collectively affect the properties of printed objects is fundamental to obtaining high performant components [15–18]. Secondly, mechanical properties are significantly affected by building orientation of samples; for this reason, some authors addressed their research toward this subject [13,19–21]. Other properties of Ultem 9085 have been only partially investigated. Cicala et al. [22] measured the rheological, morphological, and thermomechanical properties of Ultem 9085. In addition, Shelton et al. [23] focused their attention on the effect of the thermal profile of the FDM process on the inter-layer bonding of Ultem 9085 parts.

However, to the best of our knowledge, very limited literature is present on the study of thermal behavior of this material and on the effects that exposure to significant temperature and humidity variations may have on layer-by-layer 3D-manufactured parts. The promising performance of Ultem 9085 and its potential application in the transport sector, with particular reference to the automotive sector, generates interest in testing this material under severe environmental conditions.

Bagsik et al. [24] investigated the tensile properties of Ultem 9085 after long-term ageing which involved a conditioning time period up to 52 weeks and a subsequent exposure to specific temperature from −60 °C to 160 °C. These authors found that no change in geometry occurred after a long period of storage in different environmental conditions; on the contrary, the strength properties worsened when increasing test temperatures were applied.

The present paper investigates the mechanical properties of Ultem 9085 as a function of building directions. The influence on the mechanical behavior of different ageing treatments, performed by varying the environmental conditions in terms of maximum-reached temperature and relative humidity, were also studied. In addition, the thermal and thermo-oxidative behavior of this material was investigated.

2. Materials and Methods

This work focused on Ultem 9085, a high-performance thermoplastic polymer specifically designed for FDM technology; it is supplied by Stratasys in the form of natural-colour filament with a diameter of 1.75 mm. This material was used for the preparation of specimens by using a Fortus 450mc 3D-printer supplied by Stratasys. The fabrication of samples consists of three main steps. Firstly, the test coupons were modelled using the commercial software Solidworks, and then a machine code in STL (stereolithography) format was generated. Secondly, the STL file was exported into the software package Insight 12.1, which was supplied with the AM machine. It was used to set the build parameters and control all the printing stages. All the samples were fabricated by using the same building parameters: raster width of 0.508 mm, contour width of 0.508 mm, air gap of 0 mm, and contour number of 3. Moreover, filling of layers (a theoretical 100% infill was set) was performed alternating raster angles of ±45° with respect to the x-axis. Finally, the STL file was sent to FDM machine, which began to fabricate the specimens by extruding Ultem 9085 filament and depositing it layer by layer. The scanning strategy involved the deposition of the contour of a single layer followed by its filling according to the prefixed raster angle. A nozzle with a T16 tip compatible with Stratasys's equipment and showing a diameter of 0.4064 mm was used. For all other process parameters, fixed default settings supplied by Stratasys and saved in the management software were used [25].

The specimens for tensile and bending tests were manufactured according to XY, XZ, and ZX building orientations (Figure 1), where the first letter specifies the direction of the main axis of the specimen with respect to the build platform, and the second letter, together to the first one, identifies the plane on which the largest sample surface lies.

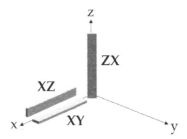

Figure 1. FDM samples built according XY, XZ, and ZX directions.

The density of both filament and printed components (with size $10 \times 10 \times 10$ mm^3) was measured by Archimedes' method using methanol as the immersion medium—the volume of displaced liquid corresponded to the samples' volume. The mass of both filament and samples was weighed by using a balance with an accuracy of 1.0×10^{-5} g. The measurements were repeated on sets of five samples for each building orientation; average values and their standard deviations were calculated.

Mechanical tests were performed using a universal testing machine (MTS Criterion Model 43, MTS Systems s.r.l., Italy) equipped with a 5 kN load cell; the configuration of the equipment was adapted to the different kinds of mechanical test. ASTM D638-14 was used as reference for the fabrication of samples and the measurement of yield strength, ultimate tensile strength, elastic modulus, and elongation at break of Ultem 9085 specimens. The average values of each property and the relevant standard deviations were calculated. Type I dog-bone samples were built according the following dimensions—overall length of 165 mm, thickness of 3.2 mm, and length and width of narrow section of 57 mm and 13 mm, respectively. The tensile tests were carried out on 15 samples (5 for each building direction) setting a strain rate of 5 mm/min in accordance to the previously-mentioned test standard; a 25 mm gauge-length extensometer was used for strain measurements.

Fracture surfaces of samples with different building orientations were observed by using Leica MS5 stereo microscope, Leica, Heidelberg equipped with Leica LAS software.

Three-point flexural tests were performed on a set of 15 specimens (5 for each building orientation) following the ASTM D790-17 standard. The results were averaged, and the standard deviation values were calculated. Rectangular bars with a length of 127 mm, a width of 12.7 mm, and 3.2 mm high were built; all the specimens were maintained at a fixed temperature of 87 °C for 12 h before performing the mechanical tests. The tests determined the flexural strength, the elastic modulus, and flexural strain of samples. A test specimen with rectangular cross section was placed in a flat position on two supports; the span length between them was fixed to 51.2 mm in accordance with the test standard. The load was then applied by means of a loading nose located in the centre of the span length; a crosshead rate of 1 mm/min was used. The specimen was deflected until rupture occurred or until a maximum strain of 9.0% was reached.

For both tensile and flexural properties average values and standard deviations were calculated. All data were statistically evaluated using one-way ANOVA. Post hoc Tukey's honestly significant difference(HSD) multiple comparison tests were then used to identify statistically-homogeneous subsets ($\alpha = 0.05$).

The thermal stability of Ultem 9085 (in the form of filament and printed samples) was evaluated through thermal-gravimetric analyses (TGA/SDTA851 Mettler Toledo), which were carried out from 25 to 800 °C with a heating rate of 10 °C/min under both argon atmosphere and air (gas flowing at 50 mL/min). The presence of any crystalline phases in the solid residue obtained after the thermal degradation of Ultem 9085 in inert atmosphere was investigated by using X-ray diffraction (Panalytical X'PERT PRO PW3040/60, Cu Kα radiation at 40 kV and 40 mA, Panalytical BV, Almelo, The Netherlands). The spectrum was collected in 2 Theta range from 10° to 80° setting a step size of 0.013°.

The ageing behavior of Ultem 9085 was evaluated after three kinds of artificial ageing treatments:

- Warm storage, which consisted of maintaining samples at a temperature of 100 °C and a zero percent relative humidity for a period of 7 days;
- Cyclic climate change, involving the constant and periodic variation of climate cell temperature between −40 and +90 °C using a thermal gradient of 1 °C/min. Both the maximum and minimum temperatures were maintained for 4 h; the relative humidity was also controlled reaching a maximum of 85% at 90 °C. This cycle was repeated for 10 times with a cumulative duration of 120 h;
- Thermal shock, consisting of two steps. Firstly, the samples were maintained at the temperature of 70 °C and zero relative humidity for 7 days. They then were subjected to a strong temperature variation up to −20 °C for 24 h.

To the best of authors' knowledge, there is not a standard procedure to determine the thermal behavior of a car component in different environmental conditions. However, some studies in the literature investigated the vehicle cabin temperatures that can be reached in various weather conditions and during the different periods of the year [26–28]. The testing conditions adopted for the three previously-described ageing treatments aim to simulate the environmental conditions (in term of temperature and relative humidity variations) that an internal vehicle component experiences during its life cycle. They are currently used for the qualification and validation of internal components in automotive industry.

Bars for flexural tests were subjected to the three kinds of previously-described environmental tests—sets of four samples were subjected to a single ageing treatment and a fourth one underwent all the treatments. The influence of the ageing treatments on the mechanical behavior of Ultem 9085 was investigated by carrying out flexural tests to samples which underwent a single ageing cycle (samples set A, after warm storage; samples set B, after cyclic climate change; and samples set C, after thermal shock) or all the three treatments (samples set D). A total of 16 specimens with building orientation XZ (which previously showed the best mechanical performance) were tested following the guidelines of ASTM D790-17 standard. The average values of flexural strength, elastic modulus, and elongation at break and their standard deviations were reported. The obtained results were statistically evaluated using one-way ANOVA. Moreover, post hoc Tukey's HSD tests were performed in order to identify the environmental conditions that significantly affected the flexural performances of Ultem 9085 ($\alpha = 0.05$).

3. Results and Discussion

3.1. Density

The measurement of density was performed on the Ultem 9085 in the form of both filament and printed components. The results are reported in Table 1.

The density of the filament was found to be equal to 1.2864 ± 0.0005 g/cm^3; this value was taken as reference to calculate the relative density of printed samples as well as the porosity values.

Table 1. Average values of densities and porosity (standard deviations in parenthesis) of printed parts for the different building orientations.

Building Orientation	Density (g/cm^3)	Relative Density (% of Theoretical)	Porosity (%)
XY	1.2389 (0.0004)	96.40 (0.03)	3.60 (0.03)
XZ	1.2552 (0.0003)	97.30 (0.02)	2.69 (0.02)
ZX	1.2637 (0.0002)	97.96 (0.02)	2.04 (0.02)

From Table 1 it is evident that samples showed increasing relative density in the printing directions XY < XZ < ZX; as expected, the corresponding values of calculated porosity showed a reverse trend. This trend of density variation when different printing directions are considered agrees with that observed by Byberg et al. [25]. However, these authors reported values of density in the range from 89.6% to 92.1%, therefore quite a lot lower than those presently obtained.

The variation of density observed for the different building orientations is mainly influenced by the ratio between the extent of the contour and the infill zones, and the scanning strategy adopted to fill each layer. The extruder head firstly deposits the contour of the object outlining its perimeter; then, it completes the layer by depositing the filament within the layer contour with an orientation of +45°/−45°. The misalignment of the infill with respect to the contour causes the formation of voids— in fact, when the +45°/−45° filament meets the contour and goes back towards the opposite side, an empty space is left at the corner of infill filament. The higher the infill area involved in the formation of each layer, the higher the fraction of formed voids.

3.2. Mechanical Properties

3.2.1. Tensile Tests

Tensile tests were performed on dog-bone specimens with building orientation XY, XZ, and ZX, respectively.

The statistical experiment was carried out using the one-way analysis of variance (ANOVA) in order to evaluate the main effects of building orientations (the independent variables) on each tensile property under investigation (yield strength, ultimate tensile strength, elastic modulus, and elongation at break, which are the response variables). Specifically, the *p*-value which indicates the level of significance of the different factors within a statistical test is reported; it represents the probability of a factor affecting the mechanical properties. The significant factors are tested with *p*-value lower than 0.05.

The statistical analysis (Table 2) revealed significant differences among building orientations for all the tensile properties under investigation; this implies that at least one of them differs from the others. The Tukey's HSD post hoc test (α = 0.05) was performed in order to identify which pairs of building directions are significantly different from each other.

Table 2. ANOVA tests for tensile properties.

Mechanical Property	Source	Sum of Squares	F	F Crit	*p*-Value
Yield strength	BG	131.961	750.954	3.885	<0.001
	WG	10.722			
Ultimate tensile strength	BG	131.961	120.109	3.885	<0.001
	WG	10.722			
Elastic modulus	BG	74,080	4.206	3.885	0,040
	WG	105,680			
Elongation at break	BG	109.256	95.608	3.885	<0.001
	WG	6.856			

BG = between groups; WG = within groups; SS = sum of squares; F = F ratio; F crit = critical F ratio.

Table 3 shows the means and the standard deviations for the tensile properties such as yield and ultimate tensile strengths, elastic modulus, and elongation at break; moreover, it compares these outcomes with those present in the technical datasheet provided by Stratasys (in this case, specimens were built by using Fortus 3D printer setting default process parameters defined by Stratasys).

Table 3. Results of tensile properties of Ultem 9085 samples built according the three building directions XY, XZ, and ZX, mean values, and standard deviations (in parentheses). Superscript letters indicate statistically-homogeneous subsets (Tukey's HSD test, α = 0:05).

Mechanical Property	XY	XZ		ZX	
	Experimental	Experimental	Datasheet	Experimental	Datasheet
Yield strength [MPa]	47.0 (0.6) [b]	54.8 (0.3) [c]	47	32.0 (1.5) [a]	33
UTS [MPa]	65.9 (0.7) [b]	73.0 (1.3) [b]	69	39.6 (6.0) [a]	42
Tensile modulus [MPa]	2220 (54) [a]	2300 (54) [a]	2150	2128 (144) [a]	2270
Elongation at break [%]	6.6 (0.6) [b]	8.0 (0.2) [b]	5.8	1.7 (0.8) [a]	2.2

It is evident that XZ orientation was the most performant one showing the highest values of yield and ultimate tensile strengths (UTS), elastic modulus, and elongation at break. However, pairwise multiple comparisons with Tukey's HSD test revealed that UTS, tensile modulus, and elongation at break of XY samples were not significantly different with respect to those determined for XZ specimens. On the other hand, ZX samples reported statistically lower tensile properties compared to the both XY and XZ building directions. Despite the value of tensile modulus for ZX samples not being significantly different from those obtained for the other two orientations, tensile strength and elongation at break were much lower with respect to those observed for XY and XZ directions. The observed trend for tensile properties is in good agreement to that reported in the literature [10,13,20,25].

A comparison between the experimental results and the mechanical properties reported in the technical datasheet (Table 3) is possible only for the best and worse performant building directions, XZ and ZX, respectively. For XZ orientation, all the measured mechanical properties were slightly higher than those stated by the manufacturer; on the contrary, this was not observed for samples printed in the ZX direction. In this case in fact, the measured properties were slightly lower than those reported in the material datasheet. Moreover, the standard deviation associated to the tensile properties of ZX samples was higher with respect to those calculated for XY and XZ samples; this implies a higher dispersion of the data.

Samples after tensile tests are shown in Figure 2a–c. These images provide evidence of the different filament deposition patterns adopted for the building of samples with XY, XZ, and ZX orientations, respectively. Figure 2d–f show the fracture surfaces of the three kinds of samples. From a visual analysis it seems that ZX cross section shows lower porosity degree than XY and XZ samples. The presence of few voids can in fact be observed mainly in the region close to the interface between infill and contour (red arrows). Similarly, in XZ samples the porosities seem mainly located at the border area where infill and contour filament meet; however, the amount of observed voids seems higher than for ZX samples. Moreover, the visual inspection of XY surface fracture evidences the presence of some voids both at the interface infill/contour and in the infill area. These qualitative observations agree to the density values reported in Table 1.

The comparison of fracture surfaces evidences some differences in the fracture behavior of samples as a function of the different building orientations.

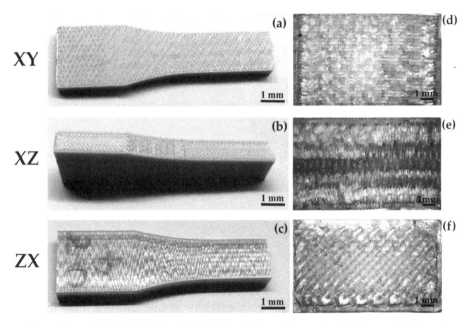

Figure 2. Samples built in XY, XZ, and ZX building directions after tensile tests. Macroscopic views of samples (**a–c**) and fracture surfaces (**d–f**).

Both XY and XZ specimens, which granted the best tensile properties, showed a brittle fracture. The slightly higher UTS of XZ samples with respect to the XY ones was probably due to the different contour/infill areas ratio. The observed fracture surfaces show a rectangular shape where most of the area is constituted by the infill, and only a minority part represents the contour. In both XY and XZ samples, the filament which forms the contours is deposited parallel to sample axes and, as a consequence, in the same direction of load application. This allows filaments to strongly oppose to the load application. On the other hand, the filaments which constitute the infill are placed at 45° with respect to the direction of load application. In samples with XZ orientation, the contour/infill areas ratio is higher with respect to XY samples because of the contour is present on the longest side of the rectangular cross section (while in the XY samples, the contour represents the shortest side). This may explain both the higher strength and elongation of XZ specimens with respect to XY ones.

In addition, the improved mechanical behavior of XZ samples with respect to XY ones can be supported by the higher degree of porosity observed in XY samples with respect to XZ ones; the porosity can in fact negatively affect the mechanical properties of the material.

The fracture surface of ZX samples is quite different from those previously discussed; it is quite flat, and the disposition of filaments at 45 °C can be clearly seen. This suggests that the fracture mechanism involves the debonding among layers, whose adhesion cannot withstand too high tensile loads. The fracture mechanism proposed for ZX samples fabricated by fused-deposition modeling was formerly reported in the literature [20,29]. The debonding at the interfaces placed perpendicular to the tensile load are also responsible for the lower strength and elongation at break.

The very good compromise between strength and ductility experimentally observed justifies the growing interest of scientific and industrial communities for Ultem 9085 material. It is worth highlighting that this material shows an ultimate tensile strength comparable to that of PA12 reinforced with carbon fibres (UTS = 76 MPa) provided by the same manufacturer [30].

3.2.2. Flexural Tests

The three-point bending tests were performed according to ASTM D790-17 standard; the statistical analysis of flexural strength and modulus results is reported in Table 4.

Table 4. ANOVA test for flexural strength and modulus.

Mechanical Property	Source	Sum of Squares	F Ratio	*p*-Value
Flexural strength	BG	3422.889	767.521	<0.001
	WG	4.460		
Elastic modulus	BG	295,285.267	208.456	<0.001
	WG	1416.533		

As previously observed for tensile properties, the *p*-values indicate that there is was a significant difference among building orientations for the flexural properties under investigation. The Tukey's HSD post hoc test was performed with the aim of identifying which building direction was significantly different from the others. Table 5 reports the values of flexural strength, elastic modulus, and flexural strain in term of means and standard deviations for each building direction.

Table 5. Flexural properties of Ultem 9085 with XY, XZ, and ZX orientations, means, and standard deviations (in parentheses). Superscript letters indicate statistically homogeneous subsets (Tukey's HSD test, $\alpha = 0.05$).

	XY	XZ		ZX	
	Experimental	Experimental	Datasheet	Experimental	Datasheet
Flexural strength [MPa]	109.72 (0.81) [b]	117.98 (0.26) [c]	112	69.98 (3.67) [a]	68
Flexural modulus [MPa]	2315 (58) [b]	2428 (17) [c]	2300	1963 (27) [a]	2050
Strain at break [%]	No break	No break	No break	3.6 (0.2)	3.7

These results confirm that samples with XZ orientation showed the best mechanical performances; in fact, the Tukey's HSD test revealed the significantly highest flexural strength and modulus for XZ samples compared to the other building directions. These values of flexural properties are slightly higher with respect to those reported in the material datasheet (these latter refer to samples built by using Fortus 3D printer setting default process parameters defined by Stratasys), as previously observed for tensile properties.

Specimens with XY orientation had significantly lower flexural properties than XZ samples. This is in good agreement with the results obtained by Byberg et al. [25] on samples printed using similar process parameters.

Moreover, according to Motaparti et al. [17] the better flexural strength of XZ coupons with respect to XY ones can be attributed to the different arrangement of contour in samples with different building orientation. When a flexural load is applied, the top surface of the specimen experiences compression, while the bottom one is under tension. In the case of XZ specimens both the top and bottom surfaces, where the stress is maximum, are mainly constituted by the contour. The presence of filaments perpendicular to the load application allow higher resistance value to be obtained. On the contrary, in XY samples the two load-bearing surfaces are mainly constituted by the infill, while the contour is only a small fraction located in the external part of the surface. This decreases the maximum load that the samples can withstand.

A significant worsening of flexural behavior was observed for ZX samples. Their flexural strength and modulus were about 40% and 20% lower than the values observed for XZ samples. Moreover, XY and XZ samples plastically deformed until the maximum value of strain was reached. On the contrary, the same tests performed on ZX samples led to the premature failure of samples at low value of strain; this provided evidence of their brittle behavior. The arrangement of interfaces among stacked layers, which are perpendicular with respect to the sample axis, and therefore parallel to the load application can be considered the main cause of the different flexural and deformation features of ZX samples.

The experimental results conclusively showed that the building direction has a significant effect on both the tensile and flexural properties. Samples with XZ orientation showed the best performances, while slightly lower mechanical properties were observed for XY specimens. On the contrary, the arrangement of stacked layers in ZX samples represents the main weakness of these samples when they are subjected to both tensile and flexural loads.

3.3. Thermal Behavior

Thanks to its good mechanical performances, Ultem 9085 material has many potential applications in the aerospace, automotive, and military industries. However, good mechanical properties are not the only requirements that a material has to satisfy for these kinds of applications. Although this material is known as flame retardant showing low smoke emission and low smoke toxicity [14], its thermal stability at high temperature and under different environmental conditions have been scarcely investigated.

3.3.1. Thermal-Gravimetric Analyses

Thermal-gravimetric analyses (TGA) were performed on the starting filament and the printed samples in order to investigate their thermal stability in a temperature range from 25 °C to 800 °C. However, no differences were observed between TGA curves of the filament and the specimens when they were independently tested in the air and argon atmospheres. This suggests that the FDM process did not influence the material thermal and thermo-oxidative degradation processes.

Figure 3 compares the TGA and derivative-TGA curves for Ultem 9085 under oxidizing (Figure 3a) and inert (Figure 3b) atmospheres.

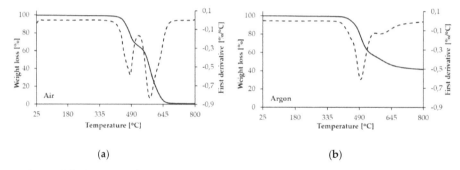

(a)   (b)

Figure 3. TGA (continuous lines) and D-TGA (dotted lines) curves for Ultem 9085 under (**a**) air and (**b**) argon atmosphere.

Both the curves collected in air and argon atmospheres show an initial degradation temperature (conventionally, it corresponds to the temperature at which there is a weight loss equals to 5%) of about 447.5 °C.

However, a significant difference in terms of solid residue could be observed—the thermo-oxidative degradation of Ultem 9085 was almost complete in air, leaving a residue of about 1%; on the contrary, a residue of about 44% with respect to the initial weight was observed after heating the material in inert atmosphere up to 800 °C. These results are comparable to those obtained by Lisa et al. [31] who investigated the thermal and thermo-oxidative stability of some polyetherimide; these authors reported a lower residue quantity when heating PEI in air (14–22 wt% at maximum temperature of 700 °C) with respect to inert atmosphere (43–54 wt%).

XRD analysis was performed on the solid residue obtained after degradation in inert atmosphere to investigate its composition. The XRD spectrum (Figure 4) shows the presence of a broad hump in 2theta range from 15° to 30° only; additional crystalline phases were not observed. This implies that the residual fraction is mainly constituted by amorphous carbonaceous species.

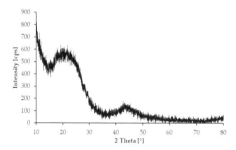

Figure 4. XRD pattern of Ultem 9085 solid residue obtained after thermal degradation.

The degradation of the material under investigation always occurred in two steps; the D-TGA curves show two peaks which correspond to the maximum rates of weight loss detected at 484.2 °C and 582.5 °C in air and at 495.8 °C and 592.5 °C in argon, respectively. The two-step degradation mechanism can be explained considering that Ultem 9085 is a mixture of PEI and polycarbonate (PC) copolymer blend; the latter is added in order to improve the material flow [13].

According to Feng et al. [32] the degradation of polycarbonate in inert atmosphere shows only one degradation step; the maximum degradation temperature is observed at 504.8 °C. On the basis of this outcome, the first degradation step of Ultem 9085 can be attributed to PC, and the second one to the degradation of PEI. A comparison between the intensity of the peak at higher temperature in

D-TGA curves is evidence of a different degradation mechanism of PEI component in air as compared to argon atmosphere.

3.3.2. Ageing Tests

The thermal behavior of Ultem 9085 was evaluated after different artificial ageing treatments:

- Warm storage, which had the aim of verifying the maintenance of high mechanical properties of the components after a long period at high temperature;
- Cyclic climate change, which allowed evaluation of the stability of the material under investigation under a constant and periodic variation of both the temperature and humidity. These variations could in fact cause reactions of hydrolytic degradation or cracks due to water penetration and freezing;
- Thermal shock, which aimed to verify the resistance of Ultem 9085 to sudden temperature variations.

In order to evaluate the effect of different ageing treatments on mechanical behavior of Ultem 9085, three-point flexural tests were performed on samples with XZ orientation (they showed the highest flexural properties, as reported in Table 5) after each ageing step (sets A, B, and C in Table 6). A set of samples was tested after the three ageing treatments (set D in Table 6).

Table 6. Tested samples after different ageing treatments.

Ageing Treatment	Set of Tested Samples	
Warm storage	A	
Cyclic climate change	B	D (A + B + C)
Thermal shock	C	

Figure 5 compares the average stress–strain curve of as-printed samples with those of specimens after different ageing treatments.

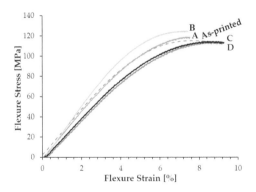

Figure 5. Stress–strain curves of samples after different ageing treatments.

The statistical analysis was performed by considering the flexural properties of Ultem 9085 samples before and after ageing treatments. The results of one-way ANOVA (Table 7) showed that the ageing factor had a significant effect on the average values of flexural strength and elastic modulus. Therefore, the Tukey's HSD post hoc test was performed with the aim of identifying which environmental conditions are significantly different from the others, mainly affecting the mechanical performances of the material.

The results in term of flexural strength, elastic modulus, and elongation at failure are reported in Table 8; these outcomes are compared to the mechanical properties of as-processed samples.

The maintenance of samples at high temperature (100 °C) for a long period in a dry environment (Figure 5, curve A) did not involve a significant variation of flexural properties with respect to those observed for as-printed specimens (Table 5). However, differently from these latter samples, samples A showed break at an elongation value close to the maximum that can be reached (as previously reported, the load was applied until the breakage of samples occurred or until a maximum strain of 9% was reached).

Table 7. ANOVA test for flexural strength and modulus obtained after ageing treatments.

Mechanical Property	Source	Sum of Squares	F Ratio	*p*-Value
Flexural strength	BG	785.737	3.914	0.021
	WG	803.0275		
Elastic modulus	BG	1,346,911	17.018	<0.001
	WG	316,574.2		

Table 8. Mechanical properties of Ultem 9085 after different ageing treatment. Superscript letters indicate statistically homogeneous subsets (Tukey's HSD test, $\alpha = 0.05$).

Mechanical Property	A	B	C	D	As Printed
Flexural strength [MPa]	118.8 (7.7) [a,b,c]	124.7 (5.3) [c]	113.4 (1.7) [a]	113.7 (13.3) [a,b,c]	118.0 (0.3) [b]
Elastic modulus [MPa]	2201(227) [b]	2650 (199) [c]	2030 (75) [a]	1956(93) [a]	2428 (17) [b]
Elongation at break [%]	7.5 (0.1)	7.5 (0.1)	No break	No break	No break

On the contrary, a significant variation of mechanical properties can be observed for samples C: the stay at the temperature of 70 °C for a long time followed by a sudden temperature variation up to −20 °C (Figure 5, curve C) leading to a significant decreasing of elastic modulus and flexural strength with respect to the as-printed samples.

Specimens that withstood cyclic climate change (Figure 5, curve B) also experienced a significant variation of temperature (the difference between minimum- and maximum-reached temperatures was 90 °C) accompanied by a change of relative humidity. The so-conditioned specimens showed a significant improvement of elastic modulus and flexural strength with respect to as-printed and aged samples. The material with increased stiffness broke at a strain value of 7.5%.

These outcomes suggest that an important thermal variation, independently from the thermal gradient, has a significant effect on Ultem 9085.

The results of Tukey's HSD post hoc test did not show a significant difference in term of elastic modulus for samples which underwent warm storage, cyclic climate change, thermal shock (Figure 5, curve D), and thermal shock only (Figure 5, curve C). This confirms that an important temperature variation has the greatest effect on mechanical behavior of the material under investigation. Moreover, it is worth nothing that the flexural strength after the three considered treatments was not significantly different from that of as printed material. Therefore, the mechanical performances of Ultem 9085 can still be considered good after the different ageing treatments. This confirms the reliability of Ultem 9085 after ageing treatments at different temperature and humidity conditions.

4. Conclusions

Fused-deposition modelling was used to process Ultem 9085, a thermoplastic polymer which has recently attracted the interest of scientific and industrial communities thanks to its good properties such as excellent flame retardancy, low smoke generation, and high mechanical performances.

In order to acquire a deeper knowledge of Ultem 9085, which is a promising candidate in many application fields, it was characterized in term of mechanical and thermal properties.

Tensile and flexural tests were performed on samples with XY, XZ, and ZX building directions. XZ orientation showed the highest yield and ultimate tensile strengths, elastic modulus, and elongation at break; however, these properties were found not to be significantly different for samples with XY

Appl. Sci. **2020**, *10*, 3170

orientation (with the exception of elastic modulus which was similar for the two kind of samples). On the contrary, ZX samples reported significantly lower tensile properties; this is due to the fact that the fracture of these samples is caused by debonding among layers, whose adhesion cannot withstand too high tensile loads. A similar trend was observed for flexural properties—XZ samples in fact showed the highest strength and modulus; however, these properties were significantly lower for both XY and ZX samples.

These outcomes confirm the excellent mechanical properties of Ultem 9085 built according XZ orientation, and provide evidence that that the building direction has a significant effect on tensile and flexural properties.

The effect of ageing treatments performed by varying the environmental conditions in term of maximum reached temperature and relative humidity on the flexural behavior of samples with XZ orientation (the most performance one) was also investigated.

Warm storage, which involves the maintenance of samples at 100 °C and zero relative humidity for a long period, did not show a significant impact on mechanical properties of Ultem 9085.

On the contrary, a sudden variation of temperature from 70 °C to −20 °C negatively affected the properties of the material under investigation, which showed a significant decreasing of both flexural strength and elastic modulus with respect to the as-printed samples. When the temperature variation was gradual, as with that experienced by Ultem 9085 during cyclic climate change, a significant improvement of flexural strength and modulus was observed.

Ultem 9085 was conclusively found to be sensitive to a sudden variation of the temperature; however, maintenance of high temperature, or a progressive variation of temperature, did not seem to significantly affect the very good mechanical performances of this material.

The thermal and thermo-oxidative behavior of this material was investigated and it was found to show no significant weight variation up to the temperature of 447.5 °C. The degradation mechanism involves two steps, which correspond to degradation of PC and PEI components, respectively.

Author Contributions: Conceptualization, G.L. and C.B.; methodology, P.C. and E.P.; software, M.G.; validation, M.G. and E.P.; investigation, M.G. and E.P.; resources, E.P. and P.C.; data curation, M.G. and E.P.; writing—original draft preparation, E.P.; writing—review and editing, E.P., M.G., and C.B.; visualization, E.P.; supervision, C.B., P.C., and G.L.; project administration, G.L. and C.B. All authors have read and agreed to the published version of the manuscript.

Funding: This research received no external funding.

Acknowledgments: The authors gratefully acknowledge the technical support provided by Raffaella Grandi, Strategic Account Manager at Stratasys and Dott. Giacomo Cacciani, Head of Additive Manufacturing Division at Overmach. The authors also thank Mario Pietroluongo who provided insight and expertise that assisted the research activities.

Conflicts of Interest: The authors declare no conflict of interest.

References

1. Boschetto, A.; Bottini, L. Robotics and Computer-Integrated Manufacturing Design for manufacturing of surfaces to improve accuracy in Fused Deposition Modeling. *Robot. Comput. Integr. Manuf.* **2016**, *37*, 103–114. [CrossRef]

2. Yang, S.; Tang, Y.; Zhao, Y.F. A new part consolidation method to embrace the design freedom of additive manufacturing. *J. Manuf. Process.* **2015**, *20*, 444–449. [CrossRef]

3. Baumers, M.; Dickens, P.; Tuck, C.; Hague, R. Technological Forecasting & Social Change The cost of additive manufacturing: Machine productivity, economies of scale and technology-push. *Technol. Forecast. Soc. Chang.* **2016**, *102*, 193–201.

4. International, A. *ASTM F2792-12a Standard Terminology for Additive Manufacturing Technologies*; ASTM International: West Conshohocken, PA, USA, 2020.

5. Equbal, A.; Islamia, J.M.; Sood, A.K.; Technology, F. Optimization of process parameters of FDM part for minimiizing its dimensional inaccuracy. *Int. J. Mech. Prod. Eng. Res. Dev.* **2017**, *7*, 57–66.

6.	Boschetto, A.; Bottini, L.; Veniali, F. Integration of FDM surface quality modeling with process design. *Addit. Manuf.* **2016**, *12*, 334–344. [CrossRef]

7.	Ngo, T.D.; Kashani, A.; Imbalzano, G.; Nguyen, K.T.Q.; Hui, D. Additive manufacturing (3D printing): A review of materials, methods, applications and challenges. *Compos. Part B* **2018**, *143*, 172–196. [CrossRef]

8.	Wang, X.; Zhao, L.; Ying, J.; Fuh, H. Effect of Porosity on Mechanical Properties of 3D Printed Polymers: Experiments and Micromechanical Modeling Based on X-ray Computed Tomography Analysis. *Polymers* **2019**, *11*, 1154. [CrossRef]

9.	Kulkarni, P.; Dutta, D. Deposition Strategies and Resulting Part Stiffnesses in Fused Deposition Modeling. *J. Manuf. Sci. Eng.* **1999**, *121*, 93–103. [CrossRef]

10.	Zaldivar, R.J.; Mclouth, T.D.; Ferrelli, G.L.; Patel, D.N.; Hopkins, A.R.; Witkin, D. Effect of initial filament moisture content on the microstructure and mechanical performance of ULTEM® 9085 3D printed parts. *Addit. Manuf.* **2018**, *24*, 457–466. [CrossRef]

11.	Jiang, S.; Liao, G.; Xu, D. Mechanical properties analysis of polyetherimide parts fabricated by fused Mechanical properties analysis of polyetherimide parts fabricated by fused deposition modeling. *High Perform. Polym.* **2018**, *31*, 97–106. [CrossRef]

12.	Geng, P.; Zhao, J.; Wu, W.; Ye, W.; Wang, Y.; Wang, S.; Zhang, S. Effects of extrusion speed and printing speed on the 3D printing stability of extruded PEEK filament. *J. Manuf. Process.* **2019**, *37*, 266–273. [CrossRef]

13.	Zaldivar, R.J.; Witkin, D.B.; Mclouth, T.; Patel, D.N.; Schmitt, K.; Nokes, J.P. Influence of processing and orientation print effects on the mechanical and thermal behavior of 3D-Printed ULTEM 9085 Material. *Addit. Manuf.* **2017**, *13*, 71–80. [CrossRef]

14.	Stratasys ULTEM™ 9085 Resin Datasheet. Available online: https://www.stratasys.com/it/materials/search/ ultem9085 (accessed on 3 February 2020).

15.	Gebisa, A.W.; Lemu, H.G. Investigating Effects of Fused-Deposition Modeling (FDM) Processing Parameters on Flexural Properties of ULTEM 9085 using Designed Experiment. *Materials* **2018**, *11*, 500. [CrossRef]

16.	Motaparti, K.P. Effect of build parameters on mechanical properties of ultem 9085 parts by fused deposition modeling. Master's Thesis, Missouri University of Science and Technology, Rolla, MO, USA, 2016. Available online: https://scholarsmine.mst.edu/masters_theses/7513 (accessed on 1 May 2020).

17.	Motaparti, K.P.; Taylor, G.; Leu, M.C.; Chandrashekhara, K.; Castle, J.; Matlack, M.; Motaparti, K.P.; Taylor, G.; Leu, M.C.; Chandrashekhara, K. Experimental investigation of effects of build parameters on flexural properties in fused deposition modelling parts. *Virtual Phys. Prototyp.* **2017**, *12*, 207–220. [CrossRef]

18.	Chuang, K.C.; Grady, J.E.; Draper, R.D. Additive manufacturing and characterization of Ultem polymers and composites. In Proceedings of the CAMX Conference Proceedings, Dallas, TX, USA, 26–29 October 2015.

19.	Bagsik, A.; Schoppner, V. Mechanical Properties of Fused Deposition Modeling Pards with Ultem* 9085. In Proceedings of the 69th Annual Technical Conference of the Society of Plastics Engineers (ANTEC'11), Boston, MA, USA, 1–5 May 2011; pp. 1294–1298.

20.	Bagsik, A.; Schoppner, V.; Klemp, E. FDM Part Quality Manufactured with Ultem * 9085. In Proceedings of the 14th International Conference Polymeric Materials 2010, Halle, Germany, 15–17 September 2010; Martin-Luther-University: Halle-Wittenberg, Germany, 2010. ISBN 9783868292824.

21.	Roberson, D.A.; Torrado, A.R.; Shemelya, C.M.; Rivera, A.; Macdonald, E.; Wicker, R.B. Comparison of stress concentrator fabrication for 3D printed polymeric izod impact test specimens. *Addit. Manuf.* **2015**, *7*, 1–11. [CrossRef]

22.	Cicala, G.; Ognibene, G.; Portuesi, S.; Blanco, I.; Rapisarda, M.; Pergolizzi, E.; Recca, G. Comparison of Ultem 9085 Used in Fused Deposition Modelling (FDM) with Polytherimide Blends. *Materials* **2018**, *11*, 285. [CrossRef] [PubMed]

23.	Shelton, T.E.; Willburn, Z.A.; Hartsfield, C.R.; Cobb, G.R.; Cerri, J.T.; Kemnitz, R.A. Effects of thermal process parameters on mechanical interlayer strength for additively manufactured Ultem 9085. *Polym. Test.* **2020**, *81*, 106255. [CrossRef]

24.	Bagsik, A.; Schöppner, V.; Klemp, E. Long-term ageing effects on fused deposition modeling parts manufactured with ultem*9085. In Proceedings of the 23rd Annual International Solid Freeform Fabrication Symposium—An Additive Manufacturing Conference, Austin, TX, USA, 22 August 2012; pp. 629–640.

25.	Byberg, K.I.; Gebisa, A.W.; Lemu, H.G. Mechanical properties of ULTEM 9085 material processed by fused deposition modeling. *Polym. Test.* **2018**, *72*, 335–347. [CrossRef]

26. Dadour, I.R.; Almanjahie, I.; Fowkes, N.D.; Keady, G.; Vijayan, K. Temperature variations in a parked vehicle. *Forensic Sci. Int.* **2011**, *207*, 205–211. [CrossRef]

27. Guard, A.; Gallagher, S.S. Heat related deaths to young children in parked cars: An analysis of 171 fatalities in the United States, 1995–2002. *Inj. Prev.* **2005**, *11*, 33–37. [CrossRef]

28. Grundstein, A.; Meentemeyer, V.; Dowd, J. Maximum vehicle cabin temperatures under different meteorological conditions. *Int. J. Biometeorol.* **2009**, *53*, 255–261. [CrossRef]

29. Badini, C.; Padovano, E.; Lambertini, V.G. Preferred orientation of chopped fibers in polymer-based composites processed by selective laser sintering and fused deposition modeling: Effects on mechanical properties. *J. Appl. Polym. Sci.* **2020**. [CrossRef]

30. Stratasys FDM Nylon 12CF FDM Nylon 12CF. Available online: https://www.stratasys.com/it/materials/search/fdm-nylon-12cf (accessed on 9 March 2020).

31. Lisa, G.; Hamciuc, C.; Hamciuc, E.; Tudorachi, N. Thermal and thermo-oxidative stability and probable degradation mechanism of some polyetherimides. *J. Anal. Appl. Pyrolysis* **2016**, *118*, 144–154. [CrossRef]

32. Feng, Y.; Wang, B.; Wang, F.; Zhao, Y.; Liu, C.; Chen, J.; Shen, C. Thermal degradation mechanism and kinetics of polycarbonate/silica nanocomposites. *Polym. Degrad. Stab.* **2014**, *107*, 129–138. [CrossRef]

 *applied sciences*

Article

Filament Extrusion and Its 3D Printing of Poly(Lactic Acid)/Poly(Styrene-*co*-Methyl Methacrylate) Blends

Luis Enrique Solorio-Rodríguez and Alejandro Vega-Rios *

Centro de Investigación en Materiales Avanzados, S.C., Miguel de Cervantes No. 120. Chihuahua C.P. 31136, Mexico; luis.solorio@cimav.edu.mx
* Correspondence: alejandro.vega@cimav.edu.mx; Tel.: +52-01-614-439-4831

Received: 18 September 2019; Accepted: 22 October 2019; Published: 28 November 2019

Abstract: Herein, we report the melt blending of amorphous poly(lactide acid) (PLA) with poly(styrene-*co*-methyl methacrylate) (poly(S-*co*-MMA)). The PLA_x/poly(S-*co*-MMA)$_y$ blends were made using amorphous PLA compositions from 50, 75, and 90wt.%, namely PLA_{50}/poly(S-*co*-MMA)$_{50}$, PLA_{75}/poly(S-*co*-MMA)$_{25}$, and PLA_{90}/poly(S-*co*-MMA)$_{10}$, respectively. The PLA_x/poly(S-*co*-MMA)$_y$ blend pellets were extruded into filaments through a prototype extruder at 195 °C. The 3D printing was done via fused deposition modeling (FDM) at the same temperature and a 40 mm/s feed rate. Furthermore, thermogravimetric curves of the PLA_x/poly(S-*co*-MMA)$_y$ blends showed slight thermal decomposition with less than 0.2% mass loss during filament extrusion and 3D printing. However, the thermal decomposition of the blends is lower when compared to amorphous PLA and poly(S-*co*-MMA). On the contrary, the PLA_x/poly(S-*co*-MMA)$_y$ blend has a higher Young's modulus (E) than amorphous PLA, and is closer to poly(S-*co*-MMA), in particular, PLA_{90}/poly(S-*co*-MMA)$_{10}$. The PLA_x/poly(S-*co*-MMA)$_y$ blends proved improved properties concerning amorphous PLA through mechanical and rheological characterization.

Keywords: amorphous poly(lactide acid); poly(styrene-*co*-methyl methacrylate); polymer blends; filament extrusion; 3D printing

---

1. Introduction

Additive manufacturing or 3D printing makes it possible to produce exceptional architecture with different complexity grades [1,2]. Additionally, additive manufacturing has several advantages, such as formability, variability, practicability, mass delivery, and surface property designs [3]. Various technologies of additive manufacturing for polymers have been developed, e.g., fused deposition modeling (FDM), bioprinting, selective laser sintering, selective heat sintering, digital light projection, and laminated object manufacturing [4,5]. The polymeric materials employed in these technologies are pellets, polymerizable resins, powders, gels, dispersed solutions, and filaments. Furthermore, the FDM limitations are the high-temperature manufacturing of polymeric filaments prior to 3D printing, exclusivity for thermoplastic polymers, and the lack of polymeric filaments available at the industrial level with mechanical properties suitable for 3D printing [1,6]. The filament extrusion conditions have a few reports in this field [7–9]. For example, Mirón et al. [8] produced uniform filaments extruded with a nozzle diameter of 2.85 mm and a temperature range from 175 to 180 °C for semi-crystalline PLA. Similarly, Kariz et al. [9] obtained PLA-wood filaments at a higher temperature (230 °C) with a nozzle of 0.4 mm. Finally, the additive manufacturing applications cover diverse areas, specifically biomedical fields such as scaffolds [10–13], drug delivery systems [14], surgical tools, and implantable devices [15,16], among others.

Moreover, PLA has been reported for additive manufacturing [16–21]. For instance, Zuniga et al. [16] replicated an amputee's finger through FDM using PLA/copper nanoparticle composites. Chacón et al. [19] printed semi-crystalline PLA under the following conditions: Temperature = 210 °C, feed rate = 20 mm/s, and a filament diameter of 1.75 mm. For polymethyl methacrylate (PMMA), Nagrath et al. [22] reported printing parameters as follows: Temperature = 275 °C, print speed = 10 mm/s, and a nozzle diameter of 0.4 mm.

Notwithstanding, PLA can be mixed with other (co)polymers, natural or synthetic, as poly(S-co-MMA), producing polymer blends with specific characteristics or properties for a given application [23]. In addition, semi-crystalline PLA, PMMA, and polystyrene (PS) have been studied, with potential applications in orthopedics, scaffolds, and tissue engineering. The semi-crystalline PLA focus has been on cell proliferation, vascularization, shape-memory, and mechanical properties [15,24–30]. PLA can be biodegraded under natural body conditions [31]. However, new scaffold architectures can be designed by delaying or anticipating the amorphous PLA biodegradation when PLA is blended with non-biodegradable polymers. In contrast, traditionally commercially available prepolymers of PMMA, PMMA-co-PS, or their mixtures, constitute the main component of acrylic bone cement [32]. Therefore, PLA_x/poly(S-co-MMA)$_y$ blends can open new opportunities for device production through additive manufacturing [15,17]. Additionally, the polymer blends containing PS, PMMA, or both have had several reports concerning miscibility and compatibility between materials [33,34]. Contrarily, few authors support PS/PMMA blend immiscibility [35,36].

In the present study, we reported the processing conditions for filament extrusion and additive manufacturing for PLA_x/poly(S-co-MMA)$_y$ blends, as well as their mechanical, thermal, and rheological properties. The key question of this research was whether a polymer blend could be made between amorphous PLA and poly(S-co-MMA) to improve on the processability of PLA during polymeric filament extrusion and 3D printing, but preserving physicochemical properties of PLA.

2. Materials and Methods

2.1. Materials

PLA Ingeo 4060D, D-lactide 12%, with an average molecular weight of 190 kg/mol, $\rho = 1.24$ g/cm^3, and glass transition temperature $T_g = 55$–60 °C from NatureWorks LLC, USA. Poly(S-co-MMA), SMMA NAS®30, $\rho = 1.090$ g/cm^3, $T_g = 103$ °C, MFI = 2.2, from Ineos Styrolution Group GmbH, Germany. In addition, the styrene and methyl methacrylate content on Poly(S-co-MMA) ranged from 70 to 90wt.% and 10 to 30wt.%, respectively [37].

2.2. Melt Blending

The amorphous PLA and poly(S-co-MMA) were dried at 60 °C for 8 h. The blending was done through a Brabender internal mixer (BB) [DDRV501, C.W. Brabender Instruments Inc., Hackensack, NJ, USA], at 50 rpm, and at a temperature of 195 °C. The blend compositions are shown in Table 1. Afterwards, the bulk sample of PLA_x/poly(S-co-MMA)$_y$ blends were ground through a blade mill to obtain PLA_x/poly(S-co-MMA)$_y$ blend pellets.

Table 1. PLA_x/poly(S-co-MMA)$_y$ blends compositions.

Samples	PLA_x/poly(S-co-MMA)y (wt.%)	
	x	y
Neat PLA	100	0
Neat poly(S-co-MMA)	0	100
PLA_{50}/poly(S-co-MMA)$_{50}$	50	50
PLA_{75}/poly(S-co-MMA)$_{25}$	75	25
PLA_{90}/poly(S-co-MMA)$_{10}$	90	10

2.3. Filament Extrusion

The filaments were produced through a single-screw extruder using $PLA_x/poly(S\text{-}co\text{-}MMA)_y$ blend pellets. A temperature of 195 °C was set. The $PLA_x/poly(S\text{-}co\text{-}MMA)_y$ blend filaments had a 1.75+/-0.1 mm average diameter. A moto-reducer with 1.6 A, 15.6 N-m of torque, and 12 V DC was used.

2.4. 3D Printing

The adequate properties for additive manufacturing were proved through a CTC 3D printer at 195 °C. The specific parameters are shown in the Table 2. The nozzle diameter was 0.4 mm.

Table 2. 3D printing parameters of $PLA_{90}/poly(S\text{-}co\text{-}MMA)_{10}$ blend.

Parameter	Value	Units
Object infill (%)	10	%
Layer Height (mm)	0.25	mm
Number of shells	3	
Feed rate (mm/s)	40	mm/s
Travel feed rate	35	
Print temperature (°C)	195	°C

2.5. Characterization

2.5.1. Mechanical Properties

The mechanical testing for $PLA_x/poly(S\text{-}co\text{-}MMA)_y$ blends, neat PLA, and neat poly(S-co-MMA) was performed using a 3382 Floor Model Universal Testing System from Instron, USA. A 5 mm/min velocity was set. The tensile mode was selected, and five probes were evaluated. The probes were manufactured in a hot press molding machine 4122 Bench Top manual press model, from Carver®, USA. The compression molding was in two stages, as follows: First, a zero load for 1 min was applied, and second, a maximum load of 2.75 tons for 3 min was used. Both stages were performed at a temperature of 225 °C. The probe's dimensions are based on ASTM D638 type IV.

2.5.2. Rheological Properties

A rotational rheometer (Physica MCR 501, Anton Paar) was used. The conditions were as follows: Oscillatory mode, parallel plate geometry with a 25 mm plate diameter, and a 1 mm gap. The analysis was achieved at a temperature of 195 °C.

2.5.3. Differential Scanning Calorimetry Analysis

Thermal studies were carried out in a DSC Q2000 differential scanning calorimeter from TA Instruments, USA. Samples of about 10 mg were sealed in standard aluminum pans under the following conditions: At a 10 °C/min heating rate and a −10–200 °C temperature range. Argon atmosphere was used in all samples.

2.5.4. Thermogravimetric Analysis

Thermal degradation was measured under air atmosphere with an SDT Q600 from TA Instruments, USA. The $PLA_x/poly(S\text{-}co\text{-}MMA)_y$ blends, neat PLA, and neat poly(S-co-MMA) were heated at a rate of 10 °C/min up to 800 °C. Approximately 16 mg of polymeric material was placed in a platinum crucible.

3. Results and Discussion

Briefly, poly(S-co-MMA) used in this research is a commercial material with random copolymer architecture. Furthermore, poly(S-co-MMA) copolymer has improved ultimate strength and elongation at break than PS, and has better thermal properties regarding PMMA. Similarly, this thermal property is one advantage over PMMA on heat transfer and melt processing [37]. In other words, this polymeric material has advantages over neat PS and PMMA.

3.1. Mechanical Properties

3.1.1. Young's Modulus

Figure 1 shows E of neat PLA, random copolymer, and the PLA$_x$/poly(S-co-MMA)$_y$ blends. The E values were 1.57, 1.58, and 1.60 GPa for PLA$_{50}$/poly(S-co-MMA)$_{50}$, PLA$_{75}$/poly(S-co-MMA)$_{25}$, and PLA$_{90}$/poly(S-co-MMA)$_{20}$ blends, respectively. The increasing E of blends can be due to compatibility between the polymer and random copolymer. The term "compatibility" refers to the blend behavior in terms of mechanical properties, and the "miscibility" is related to a homogeneous system formation at a molecular level. Usually, a miscible mixture is compatible; however, a compatible mixture is not necessarily miscible. The compatibilization of polymer blends containing PLA has been reported. For example, Quitadamo et al. [38] produced PLA/high-density polyethylene blends obtaining an optimal E = 1.88 GPa for PLA$_{50}$/HDPE$_{50}$. In another study, Balakrishnan et al. [39] improved the flexibility and E (2.2 GPa) of a PLA/low linear density polyethylene (LLDPE) blend when adding up to 15wt.% of LLDPE. Additionally, the PMMA E has been reported around 3.3 GPa, a higher value than the poly(S-co-MMA) used in the present investigation [40]. It is essential to mention that amorphous PLA has a lower E than semi-crystalline PLA, with values close to 2 GPa [31,41]. In addition, the E reported in the literature is a function of processing conditions for amorphous and semi-crystalline PLA [27]. A further example, an amorphous PLA processed (4060D from Nature Works®) at 160 °C for 10 min, in an internal mixer at 100 rpm, yielded 1.79 GPa of E [41].

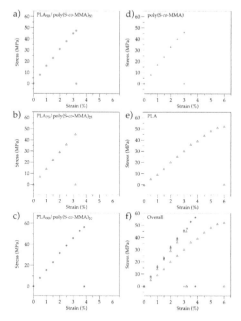

Figure 1. Young's modulus, elongation at break, and ultimate strength of (**a**) PLA$_{50}$/poly(S-co-MMA)$_{50}$; (**b**) PLA$_{75}$/poly(S-co-MMA)$_{25}$; (**c**) PLA$_{90}$/poly(S-co-MMA)$_{10}$; (**d**) neat poly(S-co-MMA); (**e**) neat PLA; and (**f**) overall.

3.1.2. Elongation at Break (%)

Contrary to E, a decreasing effect on the elongation at break was observed in the blends (Figure 1). When poly(S-*co*-MMA) content was reduced from 50 to 10wt.% in the PLA$_x$/poly(S-*co*-MMA)$_y$ blends, the elongation at break improved. The blend PLA$_{90}$/poly(S-*co*-MMA)$_{10}$ had the highest value of 3.84%. This same effect was observed on PLA/PS blends [42]. The polystyrene elongation at break was reported in other research, and it was about 4.3%, a closer value to the present result [43]. An interesting issue is that the elongation at break overrode the neat poly(S-*co*-MMA) by just adding 10wt.% of poly(S-*co*-MMA) into the PLA matrix. Finally, the poly(S-*co*-MMA) content in the blends decreases the elongation at break.

3.1.3. Ultimate Tensile Strength

The PLA$_x$/poly(S-*co*-MMA)$_y$ blends also showed compatibility in tensile strength (Figure 1). In general, it is observed that blend PLA$_{90}$/poly(S-*co*-MMA)$_{10}$ presented the highest value (56 MPa), which even surpassed the neat PLA value (52 MPa). The PLA$_{75}$/poly(S-*co*-MMA)$_{25}$ blend had a lower value than poly(S-*co*-MMA) copolymer with 45 MPa and 46 MPa, respectively.

The PLA, in some cases, can be used to increase the ultimate tensile strength for specific blends, for instance, PLA/PS blends. This improvement was attributed to low interfacial tension and high-stress transfer parameters [42]. Similarly, PLA$_{70}$/HDPE$_{30}$ blends showed an increasing effect for the ultimate tensile strength of about 49 MPa [38]. In the same manner, the PLA$_{85}$/LLDPE$_{15}$ blends displayed a maximum value of 43 MPa [39]. Table 3 displays the mechanical properties of PLA, PLA$_x$/poly(S-*co*-MMA)$_y$ blends, and poly(S-*co*-MMA).

Table 3. Mechanical properties of neat PLA, PLA$_x$/poly(S-*co*-MMA)$_y$ blends, and neat poly(S-*co*-MMA).

Sample	Young's Modulus (GPa)	Elongation at Break (%)	Ultimate Tensile Strength (MPa)
Neat PLA	1.16	6.00	52
PLA$_{50}$/poly(S-*co*-MMA)$_{50}$	1.57	3.23	48
PLA$_{75}$/poly(S-*co*-MMA)$_{25}$	1.58	3.17	45
PLA$_{90}$/poly(S-*co*-MMA)$_{10}$	1.60	3.84	56
Neat poly(S-*co*-MMA)	1.68	3.00	46

3.2. Rheological Properties

The miscibility of the binary blends was studied through the Han and Cole–Cole plot analysis because of the improvement in E concerning PLA. First, the Han plot was used to identify the polymer blend miscibility or composite materials at different temperatures and compositions [30]. The Han plots of the PLA$_x$/poly(S-*co*-MMA)$_y$ blends, neat poly(S-*co*-MMA), and neat PLA analyzed at 195 °C are displayed in Figure 2. A significant characteristic concerning the Han plot (log G' vs. log G'') is that a slope of 2 must be observed in the terminal region or low frequency if a blend is regarded as truly homogeneous [44,45]. The slope of the PLA$_{90}$/poly(S-*co*-MMA)$_{10}$ blend is the highest of all the mixtures with 1.85, while PLA$_{50}$/poly(S-*co*-MMA)$_{50}$ is the lowest, 1.62. In other words, the miscibility is present in the PLA$_x$/poly(S-*co*-MMA)$_y$ blends as a function of the PLA content obtaining adequate mechanical properties.

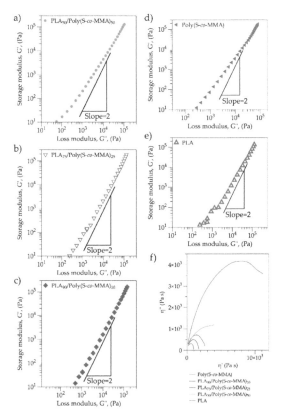

Figure 2. Han plots of (**a**) PLA$_{50}$/poly(S-*co*-MMA)$_{50}$; (**b**) PLA$_{75}$/poly(S-*co*-MMA)$_{25}$; (**c**) PLA$_{90}$/poly(S-*co*-MMA)$_{10}$; (**d**) neat poly(S-*co*-MMA); and (**e**) neat PLA; (**f**) Cole–Cole plot of neat PLA, PLA$_x$/poly(S-*co*-MMA)$_y$ blends, and neat poly(S-*co*-MMA).

Additionally, Figure 2b illustrates the Cole–Cole plots of neat PLA, neat poly(S-*co*-MMA), and PLA$_x$/poly(S-*co*-MMA)$_y$ blends. In particular, all samples present information about the relaxation process occurring in polymeric blends. These diagrams, in particular, form or become one semicircle when they indicate miscibility; on the contrary, immiscibility is attributed when more than one semicircle appears [46]. The Cole–Cole diagrams revealed the homogeneity of amorphous PLA, showing a smooth semicircular arc. A deviation from this smooth semicircular arc was observed in the case of poly(S-*co*-MMA) neat copolymer, with a more opened arc and tail. The PLA$_{50}$/poly(S-*co*-MMA)$_{50}$ blend displayed a non-closed semicircular arc accounting for immiscibility. Otherwise, the PLA$_{75}$/poly(S-*co*-MMA)$_{25}$ and PLA$_{90}$/poly(S-*co*-MMA)$_{10}$ had a similar smooth semicircular arc to the amorphous PLA, which explains the miscibility and homogeneity of these last blends [47,48]. Similarly, the Cole–Cole plots were used by Ding et al. [49] reporting two relaxation behaviors in immiscible PLA/PBAT blends, the left arc explains the polymer chain relaxation, and the right arc accounts for the droplet relaxation. Thus, Singla et al. [50] reported excellent compatibility and homogeneity for PLA/ethyl-vinyl acetate (EVA) blends for a maximum EVA content of 30wt.%. In addition, Maroufkhani et al. [51] observed a tail at the end of the curves in Cole–Cole plots, confirming phase separation between PLA and acrylonitrile butadiene rubber (NBR). In the same way, Adrar et al. [52] studied the effect of adding epoxy functionalized graphene to PLA/PBAT blends, in which a semicircular shape was observed showing positive miscibility.

Figure 3a displays the complex viscosity results, η^*, of neat PLA, neat poly(S-*co*-MMA), and PLA$_x$/poly(S-*co*-MMA)$_y$ blends. In particular, the PLA$_{75}$/poly(S-*co*-MMA)$_{25}$ and PLA$_{90}$/poly(S-*co*-MMA)$_{10}$ samples presented a broad plateau in comparison to the poly(S-*co*-MMA) copolymer. However, at high frequencies, all materials converged on similar η^* values at approximately 30 Hz. Equally important, the amorphous PLA effect on the blends is in the η^* stabilization as a frequency function. Conversely, a poly(S-*co*-MMA) disadvantage is its high η^*, and a small plateau at low frequencies, which has a shear thinning behavior. It should be noted that the PS and PMMA present a similar η^* curve compared to poly(S-*co*-MMA), although more identical to the PS caused by a more significant number of styrene monomer units according to the random copolymer composition [53].

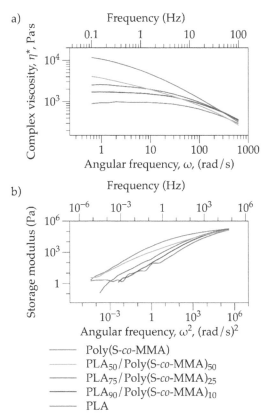

Figure 3. (**a**) Complex viscosity of neat PLA, neat poly(S-*co*-MMA), and PLA$_x$/poly(S-*co*-MMA)$_y$ blends; (**b**) Storage modulus of neat PLA, neat poly(S-*co*-MMA), and PLA$_x$/poly(S-*co*-MMA)$_y$ blends.

Additionally, the increase or decrease in η^* is the result of changes in structure when compared to starting materials. For instance, the composites present an increased η^* based on the filler content in the polymer blends [54–57]. On the other hand, in this study PLA and poly(S-*co*-MMA) presented differences in their η^* ascribable to their molecular weight, resulting in polymer mixtures with behavior between the frontier of the neat polymers. In summary, the PLA$_x$/poly(S-*co*-MMA)$_y$ blends do not show changes in structure, specifically with a physical interaction between the PLA and poly(S-*co*-MMA).

Figure 3b shows the storage module (G') graph versus square frequency of polymer blends, PLA, and poly(S-*co*-MMA). In general, PLA$_x$/poly(S-*co*-MMA)$_y$ blends converge at high frequencies; however, at low frequencies, the blends showed a predominant elastic behavior. The PLA showed the lower G' at low frequencies, while the poly(S-*co*-MMA) had a high value in the same range.

From 7 Hz to the highest frequency, PLA$_{75}$/poly(S-*co*-MMA)$_{25}$ presented a higher modulus than PLA$_{50}$/poly(S-*co*-MMA)$_{50}$. For PLA$_{90}$/poly(S-*co*-MMA)$_{10}$, there was a similar situation at approximately 30 Hz. The G′ decreased with PLA concentration in the PLA$_x$/poly(S-*co*-MMA)$_y$ blends. The G′ decreased at low frequencies according to the rheological behavior of unlinked polymers when small molecules are in the polymer structure [58].

Moreover, blend miscibility can be analyzed in the oscillatory rheology through the storage module plot versus the frequency where the behavior should be closer to a neat homopolymer. For example, this same behavior was observed in PLA-EVA [50] and PEO/PMMA blends [54]. Additionally, the PVDF/PMMA blends showed an increase in G′ compared to the neat PMMA; however, it was lower than the neat PMMA when the temperature changed [55]. Similarly, Mao et al. [56] observed an increase for G′ in the PMMA/PCE blends at low and high frequencies. These results were attributed to the short time for chain relaxation. Equally, Suresh et al. [57] reported a decrease in the rheological parameters (G′ and G″) of PVC/PMMA/rubber nitrile blends as a consequence of the flexibility provided by the rubber nitrile.

Figure 4 shows the G′ and loss modulus (G″) curves against the angular frequency, crossing point, of neat PLA, neat poly(S-*co*-MMA), and PLA$_x$/poly(S-*co*-MMA)$_y$ blends. It is well known that, at low frequencies, G″ > G′ shows a fluid state following a typical linear polymer behavior; however, at high frequencies, G′ > G″ exhibits a solid-state behavior [59–61]. Furthermore, the crossing point (G′ = G″) may change according to the composition or the branch generation due to the melt blending. For the neat PLA and poly(S-*co*-MMA), the crossing points were at 72.91 and 2.06 Hz, respectively. The PLA$_{50}$/poly(S-*co*-MMA)$_{50}$ blend presented a crossing point at 23.17 Hz, with a trend closer to the poly(S-*co*-MMA) copolymer. Subsequently, the PLA$_{75}$/poly(S-*co*-MMA)$_{25}$ crossing point was observed at 28.35 Hz. Finally, the PLA$_{90}$/poly(S-*co*-MMA)$_{10}$ had a crossing point at 57.32 Hz. Therefore, the crossing point of PLA$_x$/poly(S-*co*-MMA)$_y$ blends presented displacements, along with frequency, between amorphous PLA and poly(S-*co*-MMA) content.

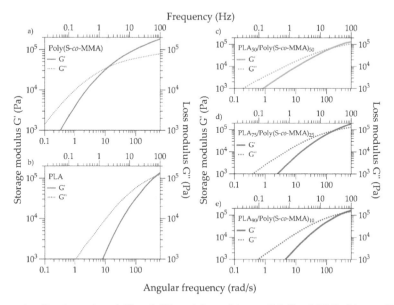

Figure 4. Crossing point of G′ and G″ modulus. (**a**) neat Poly(S-*co*-MMA); (**b**) neat PLA; (**c**) PLA$_{50}$/poly(S-*co*-MMA)$_{50}$; (**d**) PLA$_{75}$/poly(S-*co*-MMA)$_{25}$; and (**e**) PLA$_{90}$/poly(S-*co*-MMA)$_{10}$.

3.3. Thermal Analysis

3.3.1. Differential Scanning Calorimetry

The T_g dictates the miscibility phenomenon for polymer blends and processing conditions. As a rule, the polymer blend miscibility is associated with the observation of one single T_g [62]. Nonetheless, a slight T_g displacement on the PLA/PS blends suggested compatibility for those blends [42]. In other research, a single T_g for semi-crystalline PLA/PMMA blends was reported with different PMMA compositions on the PLA matrix [63]. In addition, Zhang et al. [62] evaluated the miscibility of PLA/PMMA blends through two methods, as follows: (1) Solution/precipitation, and (2) solution-casting film. The results showed that for solution/precipitation, just one T_g was observed, but two isolated T_g were present in the solution-casting film method. Figure 5 shows the curves obtained in differential scanning calorimetry, and Figure S1 displays a zoom-in of poly(S-*co*-MMA)$_y$ T_g curves belonging to PLA$_x$/poly(S-*co*-MMA)$_y$ blends. In general, all PLA$_x$/poly(S-*co*-MMA)$_y$ blends showed a T_g increasing behavior for both transition temperatures when compared with the polymer and random copolymer alone. These results suggest immiscibility, however, compatibility was observed in the E.

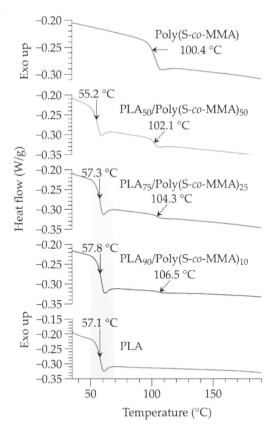

Figure 5. Heating curves (2nd cycle) of neat PLA, neat poly(S-*co*-MMA), and PLA$_x$/poly(S-*co*-MMA)$_y$ blends.

Further processing conditions (such as temperature, torque, and screw speed) for filament production are analogous to conventional extrusion. Specifically, the temperature must be between 15–60 °C above the T_g or melting point for amorphous, and semi-crystalline polymers, respectively [64].

Furthermore, these PLA$_x$/poly(S-*co*-MMA)$_y$ blends have the advantage that the system is amorphous, and therefore require less energy for their transformation. The final T$_g$ of the PLA$_x$/poly(S-*co*-MMA)$_y$ blends varied between 102.1 and 106.5 °C, which determined the melt blending temperature via extrusion at 195 °C.

3.3.2. Thermogravimetric Analysis

Figure 6 displays thermogravimetric mass loss and derivative mass loss (DTG) curves of neat PLA, neat poly(S-*co*-MMA), and PLA$_x$/poly(S-*co*-MMA)$_y$ blends. Furthermore, a particular behavior was observed in all blends presenting at least two degradation stages regarding a mass loss. For example, the PLA$_{50}$/poly(S-*co*-MMA)$_{50}$, in Figure 6a, presented a first mass loss stage of 96.8% between 321.4 and 420.8 °C, and a second step losing 3.0% over 420.8 until 490 °C. Correspondingly, there are three peaks in the DTG curve; one is located at 351 °C, followed by a second one maximum degradation temperature (T$_{max}$) = 358.6 °C, and lastly, a small peak at 488 °C. Similarly, the PLA$_{75}$/poly(S-*co*-MMA)$_{25}$ blend in Figure 6b displayed a peak with T$_{max}$ = 325.1 °C, and two shoulders placed next to the central peak with a mass loss of 16.4% from 343.6 to 400.1 °C, according to DTG and mass loss, respectively. The mass loss in the first event was about 80.8% between 255 and 343.6 °C. A third stage was also observed over 400.1 °C, with a mass loss of 1.8%. Concerning the DTG curve of the PLA$_{90}$/poly(S-*co*-MMA)$_{10}$ blend, Figure 6c shows one peak with T$_{max}$ at 334.3 °C and another at 368.5 °C. In addition, the first mass loss was observed from 250.1 to 345.6 °C, losing about 87.8%. The next stage of mass loss was of 9.4% between 345.6 and 388.6 °C. The last stage was presented at over 388.6 °C, with a mass loss of 2.0%.

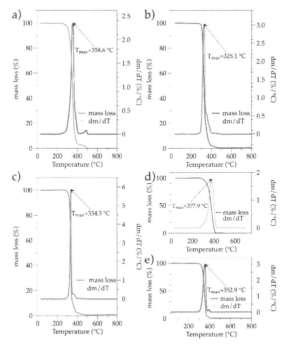

Figure 6. Thermogravimetric mass loss and DTG (dm/dT) curves of (**a**) PLA$_{50}$/poly(S-*co*-MMA)$_{50}$; (**b**) PLA$_{75}$/poly(S-*co*-MMA)$_{25}$; (**c**) PLA$_{90}$/poly(S-*co*-MMA)$_{10}$; (**d**) neat poly(S-*co*-MMA); (**e**) neat PLA.

Additionally, the degradation stages of the PLA$_x$/poly(S-*co*-MMA)$_y$ blends can be attributed to PLA degradation by hydrolysis during the process [65,66]. In addition, the chemical structure of poly(S-*co*-MMA) can decrease the thermal stability due to the presence of double carbon bonds and aromatic rings, according to Witkowski et al. [67]. Furthermore, the poly(S-*co*-MMA) had a thermal decomposition stage from 260 to 460 °C and a T_{max} around 378 °C (Figure 6d). Figure 6e displays a single degradation step for PLA with a T_{max} at 352.9 °C [50]. On the other hand, these results agreed with other publications that studied blends or composites of PLA, poly(S-*co*-MMA), PS, or PMMA. For instance, Arshad et al. [68] reported an initial decomposition temperature for poly(S-*co*-MMA) at 260 °C. In addition, Buruga et al. [69] reported three thermal decomposition stages from 360 to 474 °C for poly(S-*co*-MMA), owing to individual functional group decomposition. Likewise, it was observed that PLA/PS blends two different stages attributed to the semi-crystalline and amorphous structures for PLA and PS, respectively [70]. Similarly, Teoh et al. [71] studied the thermal decomposition of PLA/PMMA blends with or without flame retardant, finding a displacement in PMMA T_{max} from 379 °C to 430 °C when the mixture contained flame retardant. However, Mangin et al. [72] decreased the PLA T_{max} from 362 to 315 °C when incorporating phosphorus as flame retardant (5wt.%) into PLA/PMMA blends. Finally, Anakabe et al. [73] added poly(styrene-*co*-glycidyl methacrylate) P(S-*co*-GMA) to a $PLA_{80}/PMMA_{20}$ blend improving thermal stability with the copolymer at 3 pph.

Moreover, the blends' thermogravimetric curves established the conditions of thermal stability during the process of filament extrusion and 3D printing. The mass loss percentage at 195 °C for $PLA_{50}/poly(S\text{-}co\text{-}MMA)_{50}$, $PLA_{75}/poly(S\text{-}co\text{-}MMA)_{25}$, and $PLA_{90}/poly(S\text{-}co\text{-}MMA)_{10}$ were 0.03, 0.10, and 0.19%, respectively. However, poly(S-*co*-MMA) and PLA presented 0.30 and 0.02% of mass loss, respectively. Likewise, the temperatures for a mass loss at 5% for PLA and poly(S-*co*-MMA) were of 321.7 and 310.4 °C, respectively. Similarly, Cuadri et al. [74] (2018) reported a temperature of 322.4 °C for a mass loss of 5% for semi-crystalline PLA. Concerning poly(S-*co*-MMA), Zubair et al. (2017) reported a temperature of 367 °C for 5% of mass loss [75]. Regarding $PLA_x/poly(S\text{-}co\text{-}MMA)_y$ blends, the mass loss at 5% was observed at 309.2, 305.0, and 315.8 °C for $PLA_{50}/poly(S\text{-}co\text{-}MMA)_{50}$, $PLA_{75}/poly(S\text{-}co\text{-}MMA)_{25}$, and $PLA_{90}/poly(S\text{-}co\text{-}MMA)_{10}$, respectively.

3.3.3. Filament Extrusion and 3D Printing

The aim of blending poly(S-*co*-MMA) with amorphous PLA was to evaluate the material under processing conditions for filament production and additive manufacture (see Figure 7). Additionally, the methodology was developed specifically for the amorphous PLA, obtaining blends with improved properties. In general, these blends present similar E to the poly(S-*co*-MMA); however, a decrease in the elongation at break and tensile strength were observed. However, the best polymer blend was $PLA_{90}/poly(S\text{-}co\text{-}MMA)_{10}$ concerning the E.

Moreover, the three filaments of $PLA_x/poly(S\text{-}co\text{-}MMA)_y$ blends were produced via extrusion. The extruder at the prototype level was designed with two heating zones, as follows: The first zone was a pre-heating of the feed polymer blend until 150 °C, and the second had a thermal resistance in the extrusion die at 195 °C to avoid thermal degradation of the blends. The screw speed was established between 20–40 rpm as a limit for obtaining 1.75 mm filament diameter, according to the experimental values reported by Mirón et al. [8]. The filament diameter was 1.75+/−0.03 mm of standard deviation.

The $PLA_{90}/poly(S\text{-}co\text{-}MMA)_{10}$ blend was selected to show the 3D printing process in the present research. A uniform 1.75 mm diameter was reached; this size was reported by other authors [24,76]. Nevertheless, another 1.6 mm diameter was reported for PLA-hydroxyapatite filaments [21]. The printing temperature is typically set between 180–240 °C for blends and composites having PLA as the matrix. For example, in a few investigations, 210 °C were used when printing PLA [21,77], but also, a 190 °C lower temperature was set in another analysis [76].

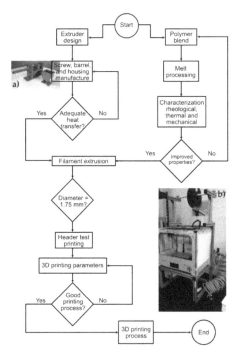

Figure 7. Scheme of the filament extrusion and 3D printing process. (**a**) Screw, barrel, and housing manufacture. (**b**) 3D printing.

The proper filament was charged to a CTC 3D printer. Parameters were set (Table 2), such as temperature and feed rate printing, a cube 1 cm × 1 cm × 1 cm for probes. The temperature variable was analyzed from 195 to 210 °C, where at 210 °C, the 3D printing piece presented melted layers with a 35 mm/s feed rate. The behavior at 200 °C was improved in the printing piece. Finally, an optimal piece was obtained at 195 °C, and at a 40 mm/s feed rate. A small 2 cm cube was used as a model (see Figure 8d). In addition, the parts manufactured for mechanical analysis through compression molding, polymeric filament via extrusion, and 3D printing by means of FDM do not undergo changes color whenever compared. Therefore, although there is slight thermal degradation in the blends, the parts do not have yellowing in the final product.

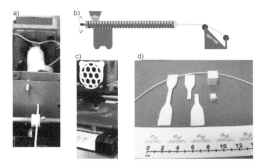

Figure 8. Filament extrusion and its 3D printing of PLA$_x$/poly(S-*co*-MMA)$_y$ blends. (**a**) Filament extrusion level prototype. (**b**) Scheme of filament extrusion. (**c**) 3D-printing. (**d**) Some parts manufactured through compression molding (probes), filament extrusion, and 3D printing, comparing their color.

4. Conclusions

The filament extrusion was obtained through a prototype using conditions resulting from thermal and rheological characterization. The thermal stability of blends at 195 °C established the final design of the prototype, placing the main heating source closer to the die. This modification also allowed filaments to be extruded without apparent changes in color at different process conditions. The filament diameter was 1.75+/−0.1 mm, according to the print head.

The 3D printing (FDM) conditions were based on the PLA_{90}/poly(S-*co*-MMA)$_{10}$ blend, as well as being confirmed with the other mixtures. The temperature and feed rate can directly influence 3D printing because it is a thermoplastic polymer. For example, at a high temperature and feed rate, the final product may have a deformation and a color change (thermal degradation). Furthermore, these blends were printed at temperatures from 210 to 195 °C, finding better conditions at 195 °C. Concerning feed rate was set at 40 mm/s.

The amorphous PLA degradation during the mixing process with poly(S-*co*-MMA) produces polymer chains with lower molecular weight than neat PLA, improving compatibility between the poly(S-*co*-MMA) and amorphous PLA. Compatibility was verified in mechanical properties because Young's modulus was improved for the PLA_{50}/poly(S-*co*-MMA)$_{50}$, PLA_{75}/poly(S-*co*-MMA)$_{25}$, and PLA_{90}/poly(S-*co*-MMA)$_{10}$ blends. The PLA_{90}/poly(S-*co*-MMA)$_{10}$ blend showed the highest value for tensile strength and elongation at break (%) due to a more significant degraded modified polymer generation. On the other hand, the complex viscosity of the blends was improved when compared with neat PLA. However, the complex viscosity of the PLA_{90}/poly(S-*co*-MMA)$_{10}$ blend showed a broad plateau lower than PLA. Furthermore, the PLA_{75}/poly(S-*co*-MMA)$_{25}$ blend showed a displacement to the left, almost reaching the neat poly(S-*co*-MMA) crossing point. All PLA_x/poly(S-*co*-MMA)$_y$ blends had two glass transition temperatures closer to the neat polymer and copolymer, suggesting compatibility in the system. The thermal property of PLA_x/poly(S-*co*-MMA)$_y$ blends showed a lower first decomposition temperature when compared with the PLA and poly(S-*co*-MMA).

Finally, mixing amorphous PLA with poly(S-co-MMA) improves processing for the filament extrusion and 3D printing, Young's modulus due to compatibility, and the complex viscosity. In addition, 3D printing of manufactured parts does not produce yellowing, although there is a slight thermal degradation.

Supplementary Materials: The following are available online at http://www.mdpi.com/2076-3417/9/23/5153/s1.

Author Contributions: Conceptualization, methodology, investigation, writing—review, and editing, L.E.S.-R. and A.V.-R.

Funding: This research received no external funding.

Acknowledgments: We wish to thank the National Council for Science and Technology of Mexico (CONACYT). We are also grateful to Erika Ivonne López Martínez, Rubén Castañeda Balderas, and Daniel Lardizábal Gutiérrez for their helpful collaboration during this research.

Conflicts of Interest: The authors declare no conflict of interest.

References

1. Wang, S.; Daelemans, L.; Fiorio, R.; Gou, M.; D'hooge, D.R.; De Clerck, K.; Cardon, L. Improving mechanical properties for extrusion-based additive manufacturing of poly(lactic acid) by annealing and blending. *Polymers* **2019**, *11*, 1529. [CrossRef] [PubMed]
2. Terekhina, S.; Skornyakov, I.; Tarasova, T.; Egorov, S. Effects of the infill density on the mechanical properties of nylon specimens made by filament fused fabrication. *Technologies* **2019**, *7*, 57. [CrossRef]
3. Henkel, J.; Woodruff, M.A.; Epari, D.R.; Steck, R.; Glatt, V.; Dickinson, I.C.; Choong, P.F.M.; Schuetz, M.A.; Hutmacher, D.W. Bone regeneration based on tissue engineering conceptions—A 21st century perspective. *Bone Res.* **2013**, *1*, 216–248. [CrossRef]
4. Liu, J.; Sun, L.; Xu, W.; Wang, Q.; Yu, S.; Sun, J. Current advances and future perspectives of 3D printing natural-derived biopolymers. *Carbohydr. Polym.* **2019**, *207*, 297–316. [CrossRef]

5. Stansbury, J.W.; Idacavage, M.J. 3D printing with polymers: Challenges among expanding options and opportunities. *Dent. Mater.* **2016**, *32*, 54–64. [CrossRef]

6. Alhnan, M.A.; Okwuosa, T.C.; Sadia, M.; Wan, K.-W.; Ahmed, W.; Arafat, B. Emergence of 3D printed dosage forms: Opportunities and challenges. *Pharm. Res.* **2016**, *33*, 1817–1832. [CrossRef]

7. Gregor-Svetec, D.; Leskovšek, M.; Brodnjak, U.V.; Elesini, U.S.; Muck, D.; Urbas, R. Characteristics of HDPE/cardboard dust 3D printable composite filaments. *J. Mater. Process. Technol.* **2020**, *276*. in press. [CrossRef]

8. Mirón, V.; Ferrándiz, S.; Juárez, D.; Mengual, A. Manufacturing and characterization of 3D printer filament using tailoring materials. *Procedia Manuf.* **2017**, *13*, 888–894. [CrossRef]

9. Kariz, M.; Sernek, M.; Obućina, M.; Kuzman, M.K. Effect of wood content in FDM filament on properties of 3D printed parts. *Mater. Today Commun.* **2018**, *14*, 135–140. [CrossRef]

10. Domingos, M.; Gloria, A.; Coelho, J.; Bartolo, P.; Ciurana, J. Three-dimensional printed bone scaffolds: The role of nano/micro-hydroxyapatite particles on the adhesion and differentiation of human mesenchymal stem cells. *Proc. Inst. Mech. Eng. Part H J. Eng. Med.* **2017**, *231*, 555–564. [CrossRef]

11. De Santis, R.; D'Amora, U.; Russo, T.; Ronca, A.; Gloria, A.; Ambrosio, L. 3D fibre deposition and stereolithography techniques for the design of multifunctional nanocomposite magnetic scaffolds. *J. Mater. Sci. Mater. Med.* **2015**, *26*, 1–9. [CrossRef] [PubMed]

12. Singh, R.; Singh, G.; Singh, J.; Kumar, R. Investigations for tensile, compressive and morphological properties of 3D printed functional prototypes of PLA-PEKK-HAp-CS. *J. Thermoplast. Compos. Mater.* **2019**, in press. [CrossRef]

13. Lee, J.; Lee, H.; Cheon, K.-H.; Park, C.; Jang, T.-S.; Kim, H.-E.; Jung, H.-D. Fabrication of poly(lactic acid)/Ti composite scaffolds with enhanced mechanical properties and biocompatibility via fused filament fabrication (FFF)–based 3D printing. *Addit. Manuf.* **2019**, *30*. in press. [CrossRef]

14. Prasad, E.; Islam, M.T.; Goodwin, D.J.; Megarry, A.J.; Halbert, G.W.; Florence, A.J.; Robertson, J. Development of a hot-melt extrusion (HME) process to produce drug loaded Affinisol™ 15LV filaments for fused filament fabrication (FFF) 3D printing. *Addit. Manuf.* **2019**, *29*. in press. [CrossRef]

15. Pucci, J.U.; Christophe, B.R.; Sisti, J.A.; Connolly, E.S. Three-dimensional printing: technologies, applications, and limitations in neurosurgery. *Biotechnol. Adv.* **2017**, *35*, 521–529. [CrossRef]

16. Zuniga, J.M. 3D printed antibacterial prostheses. *Appl. Sci.* **2018**, *8*, 1651. [CrossRef]

17. Narayanan, G.; Vernekar, V.N.; Kuyinu, E.L.; Laurencin, C.T. Poly(lactic acid)-based biomaterials for orthopaedic regenerative engineering. *Adv. Drug Deliv. Rev.* **2016**, *107*, 247–276. [CrossRef]

18. Bodaghi, M.; Damanpack, A.R.; Hu, G.F.; Liao, W.H. Large deformations of soft metamaterials fabricated by 3D printing. *Mater. Des.* **2017**, *131*, 81–91. [CrossRef]

19. Chacón, J.M.; Caminero, M.A.; García-Plaza, E.; Núñez, P.J. Additive manufacturing of PLA structures using fused deposition modelling: Effect of process parameters on mechanical properties and their optimal selection. *Mater. Des.* **2017**, *124*, 143–157. [CrossRef]

20. An, J.; Teoh, J.E.M.; Suntornnond, R.; Chua, C.K. Design and 3D printing of scaffolds and tissues. *Engineering* **2015**, *1*, 261–268. [CrossRef]

21. Senatov, F.S.; Niaza, K.V.; Zadorozhnyy, M.Y.; Maksimkin, A.V.; Kaloshkin, S.D.; Estrin, Y.Z. Mechanical properties and shape memory effect of 3D-printed PLA-based porous scaffolds. *J. Mech. Behav. Biomed. Mater.* **2016**, *57*, 139–148. [CrossRef] [PubMed]

22. Nagrath, M.; Sikora, A.; Graca, J.; Chinnici, J.L.; Rahman, S.U.; Reddy, S.G.; Ponnusamy, S.; Maddi, A.; Arany, P.R. Functionalized prosthetic interfaces using 3D printing: Generating infection-neutralizing prosthesis in dentistry. *Mater. Today Commun.* **2018**, *15*, 114–119. [CrossRef]

23. Nofar, M.; Sacligil, D.; Carreau, P.J.; Kamal, M.R.; Heuzey, M.-C. Poly(lactic acid) blends: Processing, properties and applications. *Int. J. Biol. Macromol.* **2018**, *125*, 307–360. [CrossRef]

24. Kaur, M.; Yun, T.G.; Han, S.M.; Thomas, E.L.; Kim, W.S. 3D printed stretching-dominated micro-trusses. *Mater. Des.* **2017**, *134*, 272–280. [CrossRef]

25. Farah, S.; Anderson, D.G.; Langer, R. Physical and mechanical properties of PLA, and their functions in widespread applications - A comprehensive review. *Adv. Drug Deliv. Rev.* **2016**, *107*, 367–392. [CrossRef]

26. Russias, J.; Saiz, E.; Nalla, R.K.; Gryn, K.; Ritchie, R.O.; Tomsia, A.P. Fabrication and mechanical properties of PLA/HA composites: A study of in vitro degradation. *Mater. Sci. Eng. C* **2006**, *26*, 1289–1295. [CrossRef]

27. Heidari, B.S.; Oliaei, E.; Shayesteh, H.; Davachi, S.M.; Hejazi, I.; Seyfi, J.; Bahrami, M.; Rashedi, H. Simulation of mechanical behavior and optimization of simulated injection molding process for PLA based antibacterial composite and nanocomposite bone screws using central composite design. *J. Mech. Behav. Biomed. Mater.* **2017**, *65*, 160–176. [CrossRef]

28. Michael, F.M.; Khalid, M.; Walvekar, R.; Ratnam, C.T.; Ramarad, S.; Siddiqui, H.; Hoque, M.E. Effect of nanofillers on the physico-mechanical properties of load bearing bone implants. *Mater. Sci. Eng. C* **2016**, *67*, 792–806. [CrossRef]

29. Balk, M.; Behl, M.; Wischke, C.; Zotzmann, J.; Lendlein, A. Recent advances in degradable lactide-based shape-memory polymers. *Adv. Drug Deliv. Rev.* **2016**, *107*, 136–152. [CrossRef]

30. Hao, X.; Kaschta, J.; Pan, Y.; Liu, X.; Schubert, D.W. Intermolecular cooperativity and entanglement network in a miscible PLA/PMMA blend in the presence of nanosilica. *Polymer* **2016**, *82*, 57–65. [CrossRef]

31. Middleton, J.C.; Tipton, A.J. Synthetic biodegradable polymers as orthopedic devices. *Biomaterials* **2000**, *21*, 2335–2346. [CrossRef]

32. He, Z.; Zhai, Q.; Hu, M.; Cao, C.; Wang, J.; Yang, H.; Li, B. Bone cements for percutaneous vertebroplasty and balloon kyphoplasty: Current status and future developments. *J. Orthop. Transl.* **2015**, *3*, 1–11. [CrossRef] [PubMed]

33. Chuai, C.Z.; Almdal, K.; Lyngaae-Jørgensen, J. Phase continuity and inversion in polystyrene/poly(methyl methacrylate) blends. *Polymer* **2003**, *44*, 481–493. [CrossRef]

34. Bouzid, L.; Hiadsi, S.; Bensaid, M.O.; Foudad, F.Z. Molecular dynamics simulation studies of the miscibility and thermal properties of PMMA/PS polymer blend. *Chinese J. Phys.* **2018**, *56*, 3012–3019. [CrossRef]

35. Tüzüner, Ş.; Demir, M.M. Dispersion of organophilic Ag nanoparticles in PS-PMMA blends. *Mater. Chem. Phys.* **2015**, *162*, 692–699. [CrossRef]

36. Yun, M.; Jung, N.; Yim, C.; Jeon, S. Nanomechanical thermal analysis of the effects of physical aging on glass transitions in PS/PMMA blend and PS-PMMA diblock copolymers. *Polymer* **2011**, *52*, 4136–4140. [CrossRef]

37. Cochran, T.W. Polymer Compositions Based on SMMA. EP 3068834 B1, 19 July 2017.

38. Quitadamo, A.; Massardier, V.; Santulli, C.; Valente, M. Optimization of thermoplastic blend matrix HDPE/PLA with different types and levels of coupling agents. *Materials* **2018**, *11*, 2527. [CrossRef]

39. Balakrishnan, H.; Hassan, A.; Wahit, M.U. Mechanical, thermal, and morphological properties of polylactic acid/linear low density polyethylene blends. *J. Elastomers Plast.* **2010**, *42*, 223–239. [CrossRef]

40. Montana, J.S.; Roland, S.; Richaud, E.; Miquelard-Garnier, G. Nanostructuration effect on the mechanical properties of PMMA toughened by a triblock acrylate copolymer using multilayer coextrusion. *Polymer* **2018**, *149*, 124–133. [CrossRef]

41. Nofar, M.; Tabatabaei, A.; Sojoudiasli, H.; Park, C.B.; Carreau, P.J.; Heuzey, M.C.; Kamal, M.R. Mechanical and bead foaming behavior of PLA-PBAT and PLA-PBSA blends with different morphologies. *Eur. Polym. J.* **2017**, *90*, 231–244. [CrossRef]

42. Hamad, K.; Kaseem, M.; Deri, F.; Ko, Y.G. Mechanical properties and compatibility of polylactic acid/polystyrene polymer blend. *Mater. Lett.* **2016**, *164*, 409–412. [CrossRef]

43. Li, K.; Huang, J.; Xu, D.; Zhong, Y.; Zhang, L.; Cai, J. Mechanically strong polystyrene nanocomposites by peroxide-induced grafting of styrene monomers within nanoporous cellulose gels. *Carbohydr. Polym.* **2018**, *199*, 473–481. [CrossRef] [PubMed]

44. Li, R.; Yu, W.; Zhou, C. Phase behavior and its viscoelastic responses of poly(methyl methacrylate) and poly(styrene-co-maleic anhydride) blend systems. *Polym. Bull.* **2006**, *56*, 455–466. [CrossRef]

45. Huang, Y.; Jiang, S.; Li, G.; Chen, D. Effect of fillers on the phase stability of binary polymer blends: A dynamic shear rheology study. *Acta Materialia* **2005**, *53*, 5117–5124. [CrossRef]

46. Chopra, D.; Kontopoulou, M.; Vlassopoulos, D.; Hatzikiriakos, S.G. Effect of maleic anhydride content on the rheology and phase behavior of poly(styrene-co-maleic anhydride)/poly(methyl methacrylate) blends. *Rheol. Acta* **2003**, *41*, 10–24.

47. Mohammadi, M.; Yousefi, A.A.; Ehsani, M. Thermorheological analysis of blend of high-and low-density polyethylenes. *J. Polym. Res.* **2012**, *19*, 24–29. [CrossRef]

48. Vicente-Alique, E.; Vega, J.F.; Robledo, N.; Nieto, J.; Martínez-Salazar, J. Study of the effect of the molecular architecture of the components on the melt rheological properties of polyethylene blends. *J. Polym. Res.* **2015**, *22*, 62–73. [CrossRef]

49. Ding, Y.; Feng, W.; Huang, D.; Lu, B.; Wang, P.; Wang, G. Compatibilization of immiscible PLA-based biodegradable polymer blends using amphiphilic di-block copolymers. *Eur. Polym. J.* **2019**, *118*, 45–52. [CrossRef]

50. Singla, R.K.; Zafar, M.T.; Maiti, S.N.; Ghosh, A.K. Physical blends of PLA with high vinyl acetate containing EVA and their rheological, thermo-mechanical and morphological responses. *Polym. Test.* **2017**, *63*, 398–406. [CrossRef]

51. Maroufkhani, M.; Katbab, A.A.; Liu, W.; Zhang, J. Polylactide (PLA) and acrylonitrile butadiene rubber (NBR) blends: The effect of ACN content on morphology, compatibility and mechanical properties. *Polymer* **2017**, *115*, 37–44. [CrossRef]

52. Adrar, S.; Habi, A.; Ajji, A.; Grohens, Y. Synergistic effects in epoxy functionalized graphene and modified organo-montmorillonite PLA/PBAT blends. *Appl. Clay Sci.* **2018**, *157*, 65–75. [CrossRef]

53. Nasti, G.; Gentile, G.; Cerruti, P.; Carfagna, C.; Ambrogi, V. Double percolation of multiwalled carbon nanotubes in polystyrene/polylactic acid blends. *Polymer* **2016**, *99*, 193–203. [CrossRef]

54. Zhang, C.; Liu, X.; Liu, H.; Wang, Y.; Guo, Z.; Liu, C. Multi-walled carbon nanotube in a miscible PEO/PMMA blend: Thermal and rheological behavior. *Polym. Test.* **2019**, *75*, 367–372. [CrossRef]

55. Chiu, F.C.; Yeh, S.C. Comparison of PVDF/MWNT, PMMA/MWNT, and PVDF/PMMA/MWNT nanocomposites: MWNT dispersibility and thermal and rheological properties. *Polym. Test.* **2015**, *45*, 114–123. [CrossRef]

56. Mao, Z.; Zhang, X.; Jiang, G.; Zhang, J. Fabricating sea-island structure and co-continuous structure in PMMA/ASA and PMMA/CPE blends: Correlation between impact property and phase morphology. *Polym. Test.* **2019**, *73*, 21–30. [CrossRef]

57. Suresh, S.S.; Mohanty, S.; Nayak, S.K. Effect of nitrile rubber on mechanical, thermal, rheological and flammability properties of recycled blend. *Process Saf. Environ. Prot.* **2019**, *123*, 370–378. [CrossRef]

58. Mezger, T.G. *The Rheology Handbook*, 4th ed.; Vincentz Network: Hanover, Germany, 2014; pp. 97–106.

59. Prado, B.R.; Bartoli, J.R. Synthesis and characterization of PMMA and organic modified montmorilonites nanocomposites via in situ polymerization assisted by sonication. *Appl. Clay Sci.* **2018**, *160*, 132–143. [CrossRef]

60. Potanin, A. Rheology of silica dispersions stabilized by polymers. *Colloids Surfaces A Physicochem. Eng. Asp.* **2019**, *562*, 54–60. [CrossRef]

61. Zhang, Q.; Wu, C.; Song, Y.; Zheng, Q. Rheology of fumed silica/polypropylene glycol dispersions. *Polymer* **2018**, *148*, 400–406. [CrossRef]

62. Zhang, G.; Zhang, J.; Shenguo, W.; Shen, D. Miscibility and phase structure of binary blends of polylactide and poly(methyl methacrylate). *J. Appl. Polym. Sci. Part B Polym. Phys.* **2003**, *41*, 23–30. [CrossRef]

63. Canetti, M.; Cacciamani, A.; Bertini, F. Miscible blends of polylactide and poly(methyl methacrylate): Morphology, structure, and thermal behavior. *J. Polym. Sci. Part B Polym. Phys.* **2014**, *52*, 1168–1177. [CrossRef]

64. Repka, M.A.; Shah, S.; Lu, J.; Maddineni, S.; Morott, J.; Patwardhan, K.; Mohammed, N.N. Melt extrusion: Process to product. *Expert Opin. Drug Deliv.* **2012**, *9*, 105–125. [CrossRef] [PubMed]

65. Elsawy, M.A.; Kim, K.H.; Park, J.W.; Deep, A. Hydrolytic degradation of polylactic acid (PLA) and its composites. *Renew. Sustain. Energy Rev.* **2017**, *79*, 1346–1352. [CrossRef]

66. Chow, W.S.; Lok, S.K. Thermal properties of poly(lactic acid)/organo-mont-morillonite nanocomposites. *J. Therm. Anal. Calorim.* **2009**, *95*, 627–632. [CrossRef]

67. Witkowski, A.; Stec, A.A.; Hull, R.T. Thermal decomposition of polymeric materials. In *SFPE Handbook of Fire Protection Engineering*, 5th ed.; Hurley, M.J., Gottuk, D., Hall, J.R., Harada, K., Kuligowski, E., Puchovsky, M., Torero, J., Watts, J.M., Jr., Wieczorek, C., Eds.; Springer: New York, NY, USA, 2016; pp. 53–65.

68. Arshad, M.; Masud, K.; Saeed, A. AlBr$_3$ impact on the thermal degradation of P(S-*co*-MMA): A study performed by contemporary techniques. *Iran Polym. J.* **2012**, *21*, 143–155. [CrossRef]

69. Buruga, K.; Kalathi, J.T. Synthesis of poly(styrene-*co*-methyl methacrylate) nanospheres by ultrasound-mediated pickering nanoemulsion polymerization. *J. Polym. Res.* **2019**, *26*. in press. [CrossRef]

70. Mohamed, A.; Gordon, S.H.; Biresaw, G. Poly(lactic acid)/polystyrene bioblends characterized by thermogravimetric analysis, differential scanning calorimetry, and photoacoustic infrared spectroscopy. *J. Appl. Polym. Sci.* **2007**, *106*, 1689–1696. [CrossRef]

71. Teoh, E.L.; Mariatti, M.; Chow, W.S. Thermal and flame resistant properties of poly (lactic acid)/poly (methyl methacrylate) blends containing halogen-free flame retardant. *Procedia Chem.* **2016**, *19*, 795–802. [CrossRef]

72. Mangin, R.; Vahabi, H.; Sonnier, R.; Chivas-Joly, C.; Lopez-Cuesta, J.M.; Cochez, M. Improving the resistance to hydrothermal ageing of flame-retarded PLA by incorporating miscible PMMA. *Polym. Degrad. Stab.* **2018**, *155*, 52–66. [CrossRef]

73. Anakabe, J.; Zaldua Huici, A.M.; Eceiza, A.; Arbelaiz, A. The effect of the addition of poly(styrene-*co*-glycidyl methacrylate) copolymer on the properties of polylactide/poly(methyl methacrylate) blend. *J. Appl. Polym. Sci.* **2016**, *133*, 1–10. [CrossRef]

74. Cuadri, A.A.; Martín-Alfonso, J.E. Thermal, thermo-oxidative and thermomechanical degradation of PLA: A comparative study based on rheological, chemical and thermal properties. *Polym. Degrad. Stab.* **2018**, *150*, 37–45. [CrossRef]

75. Zubair, M.; Shehzad, F.; Al-Harthi, M.A. Impact of modified graphene and microwave irradiation on thermal stability and degradation mechanism of poly (styrene-co-methyl meth acrylate). *Thermochim. Acta* **2016**, *633*, 48–55. [CrossRef]

76. León-Cabezas, M.A.; Martínez-García, A.; Varela-Gandía, F.J. Innovative functionalized monofilaments for 3D printing using fused deposition modeling for the toy industry. *Procedia Manuf.* **2017**, *13*, 738–745. [CrossRef]

77. Tian, X.; Liu, T.; Yang, C.; Wang, Q.; Li, D. Interface and performance of 3D printed continuous carbon fiber reinforced PLA composites. *Compos. Part A Appl. Sci. Manuf.* **2016**, *88*, 198–205. [CrossRef]

Article

Qualification of a Ni–Cu Alloy for the Laser Powder Bed Fusion Process (LPBF): Its Microstructure and Mechanical Properties

Iris Raffeis [1,*], Frank Adjei-Kyeremeh [1], Uwe Vroomen [1], Elmar Westhoff [2], Sebastian Bremen [3], Alexandru Hohoi [4] and Andreas Bührig-Polaczek [1]

[1] Foundry Institute, RWTH Aachen University, Intzestraße 5, 52072 Aachen, Germany; frank.kyeremeh@rwth-aachen.de (F.A.-K.); u.vroomen@gi.rwth-aachen.de (U.V.); sekretariat@gi.rwth-aachen.de (A.B.-P.)
[2] Otto Junker GmbH, Jägerhausstr. 22, 52152 Simmerath, Germany; ew@otto-junker.de
[3] Fraunhofer Institute for Laser Technology (ILT), Steinbachstraße 15, 52074 Aachen and FH Aachen University of Applied Sciences, Goethestr. 1, 52064 Aachen, Germany; bremen@fh-aachen.de
[4] Oerlikon AM GmbH, Kapellenstraße 12, 85622 Feldkirchen, Germany; alex.hohoi91@gmail.com
* Correspondence: i.raffeis@gi.rwth-aachen.de

Received: 31 March 2020; Accepted: 11 May 2020; Published: 14 May 2020

Abstract: As researchers continue to seek the expansion of the material base for additive manufacturing, there is a need to focus attention on the Ni–Cu group of alloys which conventionally has wide industrial applications. In this work, the G-NiCu30Nb casting alloy, a variant of the Monel family of alloys with Nb and high Si content is, for the first time, processed via the laser powder bed fusion process (LPBF). Being novel to the LPBF processes, optimum LPBF parameters were determined, and hardness and tensile tests were performed in as-built conditions and after heat treatment at 1000 °C. Microstructures of the as-cast and the as-built condition were compared. Highly dense samples (99.8% density) were achieved after varying hatch distance (80 µm and 140 µm) with scanning speed (550 mm/s–1500 mm/s). There was no significant difference in microhardness between varied hatch distance print sets. Microhardness of the as-built condition (247 $HV_{0.2}$) exceeded the as-cast microhardness (179 $HV_{0.2}$.). Tensile specimens built in vertical (V) and horizontal (H) orientations revealed degrees of anisotropy and were superior to conventionally reported figures. Post heat treatment increased ductility from 20% to 31% (V), as well as from 16% to 25% (H), while ultimate tensile strength (UTS) and yield strength (YS) were considerably reduced.

Keywords: additive manufacturing; LPBF; as-built; as-cast; microstructure; microhardness; tensile test; Ni–Cu alloy

---

1. Introduction

With the isomorphic nature of nickel (Ni) and copper, the Ni–Cu system results in a complete solid solution which paves way for the development of single-phase alloys over a wide range of compositions [1]. The Monel family of Ni alloys which is based on the Ni–Cu alloy system is characterized by their relatively good strength, good weldability, excellent corrosion resistance, high toughness, and fracture toughness over a wide range of temperature. As a result, they are found in many fields of application [2–5]. The single-phase Monel series is a solid solution alloy primarily hardened conventionally by cold working and typically finds its application in the chemical and petroleum industry and in food processing, as well as in the marine industry for the production of valves, pumps, heat exchangers, boiler feed heaters, pressure vessels, shafts, and fasteners, among others [2,6,7]. Comparatively, Monel is known to have better corrosion resistance than stainless steels and copper with regard to marine applications [8,9]. It does not exhibit a ductile to brittle transition (DTB) and, therefore, performs very well, even at subzero temperatures [10]. Several variations of the Monel family of alloys, such as Monel 400, Monel 401, Monel 404, Monel R-405, and Monel K-500, are commercially available [2,11,12]. G-NiCu30Nb (DIN 2.4365/9.4365) is a typical casting alloy which is modified after the Monel 400 NiCu30Fe alloy but with significant niobium (Nb) and Si content which promotes better weldability and castability, respectively [13–15]. They are typically manufactured in the as-cast state but may be soft annealed at about 900 °C depending on the wall thickness [16].

As additive manufacturing (AM) technologies continue to receive attention, various ranges of metal alloys of steel, titanium, aluminum, cobalt–chromium, gold, silver, and nickel-based superalloys continue to have their application via this generative technology [17]. AM of nickel-based alloys predominantly involved IN625, IN718, IN738LC, and Hastelloy X, whereas others like the Nimonic 263, Chromel, and MAR-M 247 were also severally investigated for the LPBF process [18]. Not much is known in published literature concerning the application of Ni–Cu-based alloy series for the AM process, especially with LPBF. Wire arc additive manufacturing (WAAM) of Monel K500 Ni–Cu alloy was investigated by Reference [19], who concluded that WAAM manufactured components had higher strength, hardness, and ductility than those hot rolled under various heat treatment conditions. Finite element modeling (FEM) of heat transfer, as well as residual stress build-up in laser metal deposition (LMD) of Monel 400, was investigated by Reference [20], while laser and plasma arc direct energy deposition (DED) fabrication of Monel 400 bimetallic structures was also carried out by Reference [21]. This present work, however, focuses on the processability, microstructure, and mechanical properties of the Ni–Cu-based G-NiCu30Nb alloy for LPBF.

2. Materials and Methods

The G-NiCu30Nb (Deutsches Institut für Normung (DIN) 2.4365) material used for this work was supplied by Otto Junker GmbH. The ingot was gas atomized by Nanoval GmbH under argon atmosphere. The powder was generally spherical, as shown in Figure 1a, but with a few satellites, whereas Figure 1b shows an etched powder particle. Particle size analysis by Nanoval GmbH had a d_{50} of 27.2 μm. The nominal compositions (wt.%) of the cast (spectral analysis) and the subsequent gas atomized powder (representative energy-dispersive spectroscopy (EDS) measurements, INCA Oxford INCA X-Sight 7426 system) are tabulated in Table 1 below, showing no significant elemental losses during the atomization. The overall carbon content in the powder was not measured; however, local enrichments were measured by EDS. The Zeiss light optical microscope (LOM) and the Zeiss Gemini Field-Emission Gun (FEG) scanning electron microscope (SEM) were used for all microscopic material characterization. Oxford Instruments Inca X-sight 7426 energy-dispersive spectroscopy (EDS) and electron backscatter diffraction (EBSD) detectors were used for all EDS and EBSD measurements, respectively, while etching was performed for 60 s (80 mL of ethanol + 40 mL of HCL+ 2 g of $CuCl_2$). Fraunhofer Institute for Laser Technology's (ILT) laser powder bed fusion (LPBF) Aconity machine (IPG Photonics Ytterbium YLR-200-SM, 200 W laser power, wavelength of 1080 nm, 90 μm beam diameter) was used for all additive printing.

Table 1. Nominal composition (wt.%) of the cast and gas atomized powder.

Material: G-NiCu30Nb	Nominal Composition (wt.%)						
	Ni	Cu	Nb	Fe	Mn	Si	Al
Cast	60.25	31.28	2.87	3.28	1.00	1.38	0.10
Gas atomized powder	59.39	29.89	3.87	3.68	0.99	1.65	0.53

Eight cubes of $10 \times 10 \times 10$ mm^3 dimensions were LPBF-printed using a bidirectional *XY* scanning strategy which turned about at 90° between each successive layer with a *Z* building direction perpendicular to the building platform. A constant laser power of 200 W and layer thickness of 30 μm were used. Two hatch distances, 80 μm and 140 μm, were chosen along with a varying scanning speed (Vs) between 550 and 1500 mm/s. Using ANSYS Pro image analysis software, the relative densities of the cubes were determined. The relative density of each cube was determined by averaging five optical measurements taken at five different positions. Microhardness was determined using the Buehler MicroMet 5104 equipment both in as-built and in post heat-treated conditions. The measurements were determined from averaging nine different measurements on the plane parallel to the building direction away from the bottom layer near the support structure using a load of 200 g (HV$_{0.2}$). The averaged microhardness of each hatch distance (80 μm and 140 μm) print set, along with their respective scanning speeds (1000–1500 mm/s and 550–850 mm/s), was compared to as-cast material. Based on the building parameters of the as-built cubes, the tensile specimens were printed using the following process parameters: laser power (Lp) = 200 W, layer thickness (Ds) = 30 μm, hatch distance (ys) = 140 μm, and scanning speed (Vs) = 850 mm/s. Tensile tests were performed at room temperature with the INSTRON 8033 tensile test machine using a crosshead speed of 0.3 mm/min. Six tensile test specimens according to DIN 50125 standard were as-printed and tensile tested (three vertical and three horizontal orientations), while six other as-built tensile specimens also printed according to DIN 50125 standard were post heat treated (PHT) at 1000 °C for 1 h (three vertical and three horizontal orientations) before the tensile test was performed. The tensile test results obtained from each building orientation were averaged and compared.

Figure 1. SEM micrographs: (**a**) morphology of the gas atomized G-NiCu30Nb (DIN 2.4365) powder; (**b**) etched powder particle revealing dendrite structures.

A full factorial experimental design approach was adopted in this study to investigate the effect of varying hatch distances with varying scanning speeds on the printing of as-built samples, on the relative density, and on the subsequent effect of PHT on mechanical properties. The scope of the experiment also covered an investigation into the anisotropic microstructure of different building directions and the impact on mechanical properties.

3. Results

3.1. As-Built LPBF Micrographs

$$E_v = \frac{L_P}{D_s \cdot v_s \cdot y_s} \left[J / mm^3 \right]. \tag{1}$$

The volumetric energy density (E$_v$) equation as expressed in Equation (1) above highlights the inter-relation between the key process parameters such as the laser power (L$_p$), layer thickness (D$_s$), scanning speed (V$_s$), and hatch distance (y$_s$). This served as the basis for the parametric design employed in this study. The as-printed cubes were generally dense (≥99.6%) with the most dense cubes recording a relative density of 99.8%, as shown in Figure 2. Table 2 below highlights the process parameters and the achieved relative densities in each print set.

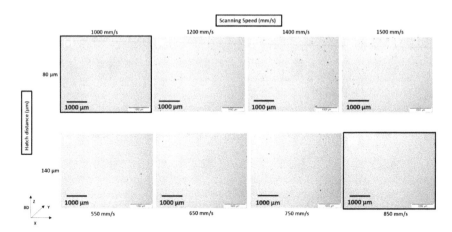

Figure 2. Light optical microscope (LOM) micrographs showing two different hatch distances with scanning speed variation at constant laser power (200 W) and layer thickness (30 μm) for unetched as-built cubes; the highest relative density of 99.8% was recorded at 1000 mm/s and 850 mm/s.

Table 2. Process parameters and the respective relative densities of the as-built cubes shown in Figure 2.

Sample	Laser Power (W)	Layer Thickness (μm)	Scanning Speed (mm/s)	Hatch Distance (μm)	Relative Density (%)
2a	200	30	1000	80	99.80
2b	200	30	1200	80	99.72
2c	200	30	1400	80	99.63
2d	200	30	1500	80	99.70
2e	200	30	550	140	99.78
2f	200	30	650	140	99.77
2g	200	30	750	140	99.71
2h	200	30	850	140	99.80

For the 80 μm hatch distance print set, the least and most dense cubes recorded relative densities of 99.63% and 99. 80%, respectively. These represented percentage porosities of 0.37% and 0.20%, respectively. Similarly, in the 140 μm hatch distance print set, 99.71% and 99.80% were recorded as the lowest and highest relative densities, respectively. This also accounted for percentage porosities of 0.29% and 0.20%, respectively. From Figure 2, for both hatch distance print sets (80 μm and 140 μm), small spherical pores can be observed, which are characteristically gas-induced pores which may have been trapped during melting and solidification [22]. The average pore size measured with the ANSYS Pro image analysis software among all the samples was less than 1 μm. In Figure 3, micrographs of 80 μm hatch distance samples at different scanning speeds are compared. The as-built cube printed at 1000 mm/s scanning speed shown in Figure 3a is observed to be denser and crack-free compared to Figure 3d, printed at 1500 mm/s scanning speed, which shows more cracks. Calculated energy density values ranged from 55–88 J/mm^3.

Since the volumetric energy density (Ev) is the same for the samples shown in Figure 2d,h (about 56 J/ mm^3), and the cracks do not resemble Lack of Fusion (LOF) defects, they are believed to be residual stress propagating cracks arising from the build-up of large amounts of local residual stress in the as-built condition due to the generally high temperature gradient associated with the LPBF process. High scanning speeds tend to generate higher cooling rates [23], which, in combination with smaller hatching distance, can lead to local stresses and fracture when the material ultimate tensile strength is exceeded [24]. The increase in scanning speed to 1500 mm/s is, therefore, believed to have induced higher cooling rates which influenced residual stress build-up and led to fracture or cracks as seen in Figure 3e,f and not in Figure 3a–c at 1000 mm/s scanning speed.

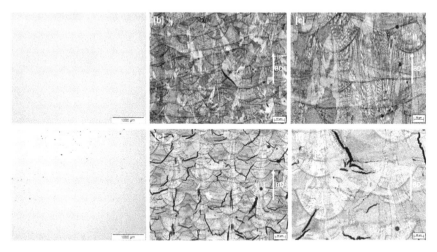

Figure 3. (**a–c**) Dense 80 μm hatch distance cube (etched and unetched) built at 1000 mm/s; (**d–f**) 80 μm hatch distance cube (etched and unetched) built at 1500 mm/s showing a high number of cracks.

The micrographs in Figure 4 show etched and unetched cubes at different magnifications built using a 140 μm hatch distance. Although there are few gas pores in these cubes, no cracks are observed with scanning speed between 550 mm/s and 850 mm/s. Lower hatch distances tend to reduce the building rate, and they are typically characterized by increased melt pool overlap, as well as an increase in local melt pool temperature (vice versa for increased hatch distance) [25]. However, for complete fusion and good metallurgical bonding, the amount of laser and material interaction time should be enough to generate the right volumetric energy density available for complete melting [26]. As previously explained, in spite of the higher hatch distance of 140 μm, as seen in Figure 4, cracks which were not observed in this print set can be attributed to the lower scanning speed between 550 and 850 mm/s. This may have led to a reduction in the build-up of high residual stresses leading to fracture or cracks as seen in Figure 3e,f.

In both hatch distance print sets (80 μm and 140 μm), however, well-defined melt pool boundaries with orientation along the building direction and a lot of epitaxial growth were observed.

Scanning electron micrographs of the as-built cubes shown in Figure 5 highlight discontinuous cellular dendritic sub-structures with high levels of epitaxial growth at different portions of the microstructure. Epitaxial growth is seen to be interrupted at some point at the top of the melt pool as shown in Figure 5a and grows again only to be terminated at the melt pool boundary. In Figure 5b, epitaxial growth is seen stretching across the melt pool boundary. In Figure 5d, it is interlaced between dendritic cellular substructures. The microstructural features, such as the growth of epitaxy and the formation of fine dendritic structures, as seen in Figure 5, are consistent with typical as-built LPBF microstructures reported in several nickel-based alloys such as IN718, IN625, and Haynes® 230® [27,28]. The unique microstructure observed in the LPBF built nickel-based alloys, which is also seen in Figure 5, is influenced not only by the process parameters, but also by the extremely fast cooling rate associated with the LPBF process, which makes it different from the cast [29].

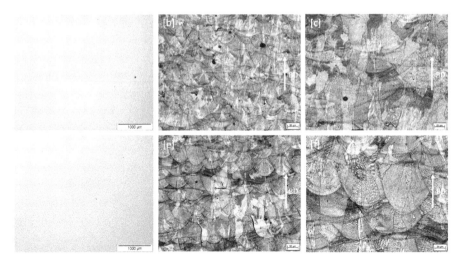

Figure 4. LOM micrographs of 140 μm hatch distance print set: (**a**) unetched, dense as-built cube (scanning speed 550 mm/s); (**b,c**) etched with CuCl₂ showing epitaxial growth and columnar grain formation; (**d**) unetched, dense as-built cube (scanning speed 850 mm/s); (**e,f**) etched with CuCl₂ also showing epitaxial growth and columnar grain formation.

3.2. As-Cast Micrographs

As expected, both LOM and SEM micrographs of the as-cast structure show a completely different microstructure compared to the as-built microstructure. Micrographs of an unetched as-cast sample in Figure 6a,b show a random distribution of coarse particles of varying sizes and shapes in the matrix. The size of these particles ranges from as small as 5 μm to as huge as 20 μm. SEM/EDS mapping and quantitative analyses are reported in both Figures 6c and 7. EBSD phase analyses confirm these particles as niobium carbide particles (NbC).

While some authors [30,31] reported the formation of NbC particles in conventionally fabricated Monel 400 Ni–Cu alloys, others [32–34] found these particles in Selective Laser Melting (SLM) or Laser Metal Deposition (LMD) built IN718 and IN625. To the best of our knowledge, no work is currently reported concerning their formation in the G-NiCu30Nb Ni–Cu alloy processed via LPBF additive manufacturing technique. Since the various magnifications employed in the characterization of the as-built microstructures could not confirm NbC in the as-built condition, further work will be carried out in the next phase of this study not only to ascertain expected NbC particles formation but also the nature of their distribution in the as-built condition.

3.3. EBSD Measurement of As-Built Samples

Inverse pole figure (IPF) EBSD maps obtained from the as-built samples are shown in Figure 8 (ys = 80 μm, Vs = 1000 mm/s and ys = 140 μm, Vs = 850 mm/s). From Figure 8, both as-built samples show columnar orientation along the building direction. The effect of constitutional supercooling is dependent on the quotient (G/R), where (G) is the temperature gradient and (R) is the rate of solidification. (G/R) determines the solidification morphology. For rapid solidification processes, like LPBF (10^4–10^8 K/s), heat is conducted away from the building plate in a directional manner. This results in the formation of typically columnar oriented structures due to the direction of flow of heat flux as shown by the EBSD measurements [26,35–37]. A melt pool exposed to the atmosphere transfers heat flux by radiation while the underlying solidified substrate does so by conduction; grain growth is, therefore, favored in the direction perpendicular to the solid–liquid front [38].

Appl. Sci. **2020**, *10*, 3401

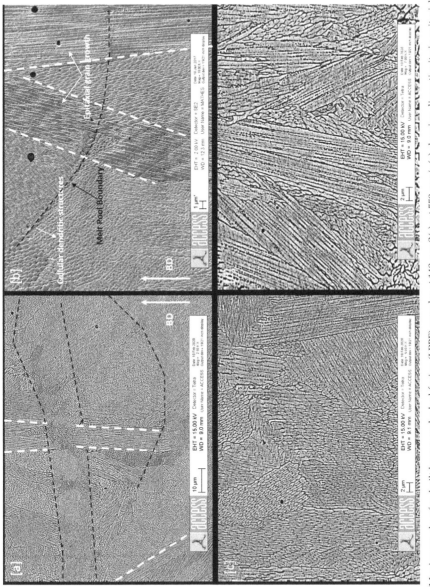

Figure 5. SEM micrographs of as-built laser powder bed fusion (LPBF) samples at 140 μm (Ys): at 550 mm/s (**a**) showing discontinuity in epitaxial growth; (**b**) combination of both epitaxial and cellular dendritic structures; at 850 mm/s (**c**), (**d**) different subgrain orientations at higher magnification.

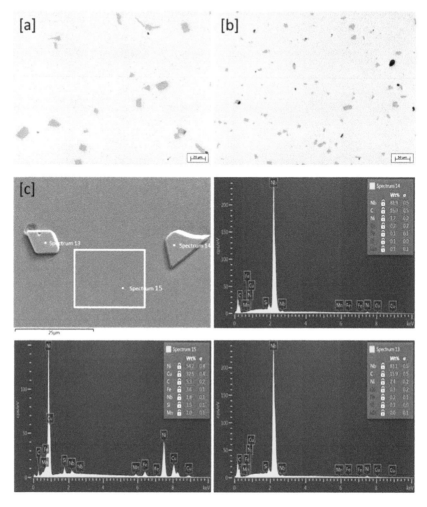

Figure 6. LOM: (**a**,**b**) LOM of unetched as-cast samples showing randomly dispersed NbC particles; SEM/energy-dispersive spectroscopy (EDS): (**c**) EDS quantitative measurements of NbC particles in as-cast sample.

The 80 μm hatch distance sample recorded an average equivalent grain diameter (ECD) of 8.45 μm with a mean aspect ratio of 2.13 compared to an average ECD of 11.93 μm with a mean aspect ratio of 2.52 for the 140 μm hatch distance as-built sample. The smaller ECD of the 80 μm hatch distance as-built samples compared to the 140 μm hatch distance samples is a result of the different local thermal and solidification behavior of the different hatch distances. The hatch distance affects the nature of local heat transfer such that wider hatch distances not only reduce the rate of local heat accumulation but also the local cooling rates and lead to coarsened microstructures [38].

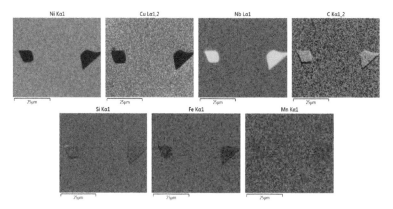

Figure 7. EDS mapping showing coarse NbC particles in an as-cast sample.

Figure 8. Inverse pole figures of as-built samples: (**a**) 80 μm hatch distance; (**b**) 140 μm hatch distance showing columnar orientation along the building direction.

3.4. Mechanical Properties

3.4.1. Microhardness

The microhardness of both as-built print sets (80 μm and 140 μm hatch distance) and that of as-cast sets were determined and compared. As illustrated by the bar chart in Figure 9, hatch distance variation did not significantly influence the microhardness of the as-built print sets as there was only a marginal difference in microhardness between the 80 μm (247 $HV_{0.2}$) and 140 μm (244 $HV_{0.2}$) hatch distance. However, the microhardness of both as-built print sets was far higher than the measured microhardness of as-cast (179 $HV_{0.2}$). The higher microhardness recorded in the as-built print sets can be attributed to a combination of both the residual stresses build-up in the as-built condition and the finer microstructures of LPBF printed samples which enhance mechanical properties compared to conventional manufactured methods like casting [26,39]. Post heat treatment (PHT) performed on the as-built samples at 1000 °C for one hour saw the microhardness significantly reduced to 221 $HV_{0.2}$ and 210 $HV_{0.2}$ for both as-built print sets with 80 μm and 140 μm hatch distance. The loss in microhardness may be associated with the relief of stresses and the coarsening of the microstructure through PHT. Even after PHT, the LPBF print set recorded higher microhardness than the as-cast material. Figure 9 and Table 3 show a comparison of the microhardness at different printing conditions.

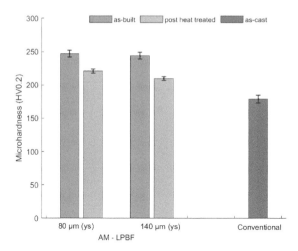

Figure 9. Microhardness of as-built, post heat treated (PHT), and as-cast material compared.

Table 3. Comparison of microhardness of as-cast, as-built, and as-built + PHT cubes.

Material: G-NiCu30Nb	$HV_{0.2}$
As-Cast	179
LPBF as-built cube (ys =80 μm)	247
LPBF as-built + PHT (ys = 80 μm)	221
LPBF as-built cube (ys = 140 μm)	244
LPBF as-built + PHT (ys = 140 μm)	210

3.4.2. Tensile Test

Tensile tests, conducted at room temperature using a crosshead speed of 0.3 mm/min on six as-built tensile specimens (three built in the vertical and three built in the horizontal orientation) showed the existence of anisotropy in the different building orientations. From Figure 10, it was observed that there was only a marginal difference in the ultimate tensile strength (UTS) and the yield strength (YS) between specimens built in the vertical orientation and specimens built in the horizontal orientation. UTS and YS of the vertical orientation-built specimen were 739 MPa and 531 MPa, respectively, compared to 731 MPa and 567 MPa of horizontal orientation specimen. However, a significant difference in the percent elongation between specimens built in both orientations was observed in Figure 9. The vertical orientation-built specimen had a significantly higher elongation of 20% compared to 16% in the horizontal orientation-built specimen. The LPBF as-built specimens were far superior in UTS and YS compared to UTS (450 MPa) and YS (170 MPa) of as-cast material [40]. The elongation of both as-built specimens (vertical and horizontal orientation) fell short of elongation of the as-cast material (25%) [40]. From Table 4, post heat treatment of tensile specimens at 1000 °C for 1 h showed a deterioration of strength properties of the alloy. PHT specimens built in the vertical orientation recorded a UTS of 674 MPa, a 9% reduction compared to the as-built state. The YS also deteriorated by 23% (408 MPa) compared to the as-built state. Similarly, for horizontally built specimens, post heat treatment at 1000 °C for 1 h recorded a UTS of 665 MPa and YS of 417 MPa, representing a reduction of 9% and 26%, respectively, as compared to the as-built state.

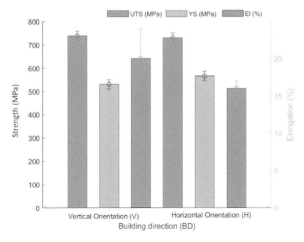

Figure 10. Tensile properties of as-built specimens (ys: 140 μm, Vs: 850 mm/s) shown in both horizontal and vertical orientations by stress–strain diagram and bar chart highlighting anisotropy in different building orientations.

Table 4. Summary of compared tensile properties.

Material: G-NiCu30Nb			
	UTS (MPa)	YS (MPa)	El (%)
As-Cast [16]	450	170	25
LPBF as-built (V) at r.t.p	739	531	20
LPBF as-built (H) at r.t.p	731	567	16
LPBF as-built + PHT (V)	674	408	31
LPBF as-built + PHT (H)	665	417	25

As-built process parameters (tensile specimens): laser power (Lp)—200 W, layer thickness (Ds)—30 μm, hatch distance (ys)—140 μm, scanning speed (Vs)—850 mm/s; r.t.p—room temperature; PHT—post heat treatment at 1000 °C, 1 h; V—vertical orientation-built specimen; H—horizontal orientation-built specimen.

4. Discussion

The attainment of a high relative density in LPBF depends on the ability to obtain the right process parameter combinations. The 80 μm hatch distance varied with scanning speed yielded a high relative density of 99.8% at 1000 mm/s scanning speed. However, when scanning speed increased to 1500 mm/s, cracks became pronounced. The cracks which are believed to be residual stress-induced cracks were, however, not seen when the hatch distance was increased to 140 μm and the scanning speed was comparatively reduced (550–850 mm/s). The role of scanning speed, as studied by various authors [23,26,41], is critical not only in determining the amount of volumetric energy density available for complete melting but also the magnitude of local residual stress build-up. Overly increased scanning speed results in insufficient interaction between laser beam and material because high scanning speed leads to reduced energy density available for complete melting, which would lead to lack of fusion defects. High scanning speeds also lead to a faster cooling rate which may generate cracks due to the high residual stress build-up as already explained. This is believed to have resulted in the cracks seen in Figure 3e,f.

No such defect was seen with an increase in hatch distance to 140 μm. This may be due to the comparatively lower scanning speed (550–850 mm/s) regime which gave sufficient beam–material interaction time to enhance good metallurgical bonding, not only between tracks but also with underlying layers. The lower scanning speed regime (550–850 mm/s) at the higher hatch distance (140 μm) may have generated considerable local residual stress which was low enough to avoid crack formation. The building of a highly dense sample of 99.8% relative density with a scanning speed of 850 mm/s was, therefore, possible at 140 μm hatch distance. Small spherical gas-induced porosity, as observed in both 80 μm and 140 μm hatch distances and described by various authors [22,42], takes its origin from the powder feedstock and from the in-process escape of elements and materials, which results in entrapped gases in the molten pool carried through to the as-built part. The elimination of this porosity can be achieved through the combination of the use of metallic powders of good packing density, good pre-process powder preparation, and highly optimized process parameters [43,44].

Samples of both as-built 80 μm and 140 μm hatch distance showed typical LPBF microstructures with defined melt pools oriented along the building direction. High epitaxial grain growth was observed in both microstructures. Epitaxial grain growth is formed from remelted areas as a result of multiple laser scans or the remelting of previous layers [44,45]. The LPBF technique generates fine microstructures owing to the high cooling rates. ECD grain size measurements by EBSD confirmed fine microstructures of ECD of 8.45 μm and 11.93 μm for the 80 μm and 140 μm hatch distance print sets, respectively. Higher hatch distances tend to generate slower cooling rates compared to smaller hatch distances which influences the grain morphology [46]. The grain size of the 140 μm hatch distance was, therefore, slightly bigger than the 80 μm hatch distance print set. As a result of the slow cooling rates of a typical conventional manufacturing process like casting, in the as-cast microstructure, undissolved and coarse NbC particles were detected with EDS/EBSD analysis, which were not detected in the as-built LPBF microstructure.

As expected, mechanical properties of as-built ($HV_{0.2}$, UTS, YS) and PHT ($HV_{0.2}$, UTS, YS, El) samples were generally superior to properties of the as-cast condition. The high microhardness observed in the as-built cubes, which is attributed to both the fine microstructure and residual stress build-up, was slightly reduced after a PHT at 1000 °C for 1 h mainly because of stress relief. The comparatively higher cooling rate of the 80 μm print set which accounted for its finer microstructure resulted in the slightly higher microhardness than in the 140 μm hatch distance print set. Anisotropy in LPBF built samples was severely investigated [39,47]; this was reflected in the tensile properties which were built in different orientations. The vertical orientation-built tensile specimen showed a higher UTS (739 MPa) and ductility (20 %), but lower YS (531 MPa). The horizontal orientation-built tensile specimen had a higher YS (567 MPa), but lower UTS (731 MPa) and El (16 %). The PHT tensile specimen instead had higher ductility in both orientations (V: El (31%) and H: El (25%)) than the as-built specimen but lower UTS (V—674 MPa, H—665 MPa) and YS (V—674 MPa, H—665 MPa). Due to the steep temperature gradient associated with the LPBF process, grain growth typically takes place in a directional manner usually along the building direction. The resulting microstructure and texture influence the mechanical anisotropy [48]. The disparity in the tensile properties may, therefore, have originated on the angle between the columnar oriented grains which were along the building direction and the tensile axis [49]. Stress relief through post heat treatment at 1000 °C for 60 min also significantly contributed to the increase in ductility at the expense of both yield and ultimate tensile stress when the as-built samples were heat-treated.

5. Conclusions

In this work, the LPBF processability of the nickel–copper-based alloy (G-NiCu30Nb/DIN 2.4365) was successfully demonstrated. After varying process parameters between scanning speed and hatch distance, highly dense cubes of relative density of 99.8% were built at both 80 μm and 140 μm hatch distances with scanning speeds of 1000 mm/s and 850 mm/s, respectively. No significant difference in microhardness was recorded for both as-built hatch distance print sets (247 $HV_{0.2}$ for ys = 80 μm and 244 $HV_{0.2}$ for ys = 140 μm). This was significantly higher than the as-cast microhardness of 179 $HV_{0.2}$ of the same material. PHT resulted in a further drop in microhardness of the as-built cubes to 221 $HV_{0.2}$ (80 μm ys) and 210 $HV_{0.2}$ (140 μm ys) after 1000 °C for 1 h. Tensile properties of both vertical and horizontal orientation-built tensile specimens showed superior properties to as-cast specimens (UTS 450 MPa, YS 170 MPa, El 25%). The UTS of vertical orientation-built specimen was higher (739 MPa) than that of the horizontal orientation-built specimen (731 MPa) but recorded a lower YS (531 MPa) compared to the horizontal orientation-built specimen (567 MPa). Similarly, anisotropy was observed with regard to the elongation and the building orientation. Elongation was observed to be higher (20%) in the vertical orientation-built specimen than in the horizontal orientation-built specimen (16%). PHT deteriorated both the UTS (674 MPa—V orientation, 665 MPa—H orientation) and YS (408 MPa—V orientation, 417 MPa—H orientation) compared to the as-built condition but also recorded a higher elongation (31%—V orientation, 25% MPa—H orientation).

Epitaxial growth does not require sufficient nucleation energy for nucleus formation. In additive manufacturing, the growth of epitaxy is associated with remelted areas which typically arise from overlapping melt pools. This phenomenon which also occurs in single-phase alloys leads to the formation of columnar structures which introduces anisotropic properties. Columnar-oriented structures are by their nature coarser than equiaxed structures and may not be fit for use in multipronged stress applications compared to a preferred equiaxed or combination of columnar-equiaxed structures [50]. Further work, therefore, needs to be done to minimize the level of epitaxy to achieve equiaxed or columnar-equiaxed structure necessary for the production of reproducible AM parts with isotropic properties. This could be attained through the introduction of heterogeneous nucleation sites which act as grain refiners and the slowing of solidification speed through in situ heat treatment techniques for nuclei formation.

After successful LPBF processability, the G-NiCu30Nb alloy proved to be a good material candidate for AM fabricated components but, however, requires further research to understand and optimize the microstructure for enhanced mechanical properties. This study significantly opens the door for industrial applications of additively manufactured parts from the G-NiCu30Nb alloy after demonstrating successful LPBF processability. This means that not only does the G-NiCu30Nb alloy add up to the scope of AM applicable materials, but also that this Monel 400 variant, which according to this study shows very good mechanical properties via LPBF processing, presents itself as an alternative to the cast in the building of convoluted or complex geometrical parts and in the elimination of tooling cost.

Author Contributions: Conceptualization, I.R.; data curation, F.A.-K. and A.H.; investigation, F.A.-K. and A.H.; methodology, I.R.; project administration, I.R.; resources, E.W. and S.B.; supervision, I.R.; validation, U.V.; writing—original draft, I.R. and F.A.-K.; writing—review and editing, I.R., U.V., A.H., and A.B.-P. All authors have read and agreed to the published version of the manuscript.

Funding: This research received no external funding.

Conflicts of Interest: The authors declare no conflicts of interest.

References

1. Lippold, J.C.; Kisser, S.D.; Dupont, J.N. *Welding Metallurgy and Weldability of Nickel-Base Alloys*; John Wiley & Sons: Hoboken, NJ, USA, 2011; p. 456.
2. Esgin, U.; Özyürek, D.; Kaya, H. An investigation of wear behaviors of different Monel alloys produced by powder metallurgy. *AIP Conf. Proc.* **2016**, *1727*. [CrossRef]
3. Xiaoping, Z.; Wenzhang, C. Experimental Research on Heating Performance of Monel K-500 Alloy. *Adv. Mater. Res.* **2012**, *572*, 273–277. [CrossRef]
4. Kutz, M. *Handbook of Materials Selection*; John Wiley & Sons, Inc.: New York, NY, USA, 2002; pp. 237–241.
5. Barsanescu, P.-D.; Leitoiu, B.; Goanta, V.; Cantemir, D.; Gherasim, S. Reduction of Residual Stresses Induced by Welding in Monel Alloy, Using Parallel Heat Welding. *Int. J. Acad.* **2011**, *3*, 335–339.
6. Neenu, J.; Jobil, V. Study on Behavior of Steel, Monel and Inconel at Elevated Temperature. *IJEDR* **2015**, *3*, 327–331.
7. Devendranath, R.K.; Arivazhagan, N.; Narayanan, S. Effect of filler materials on the performance of gas tungsten arc welded AISI 304 and Monel 400. *Mater. Des.* **2012**, *40*, 70–79. [CrossRef]
8. Devendranath, R.K.; Arivazhagan, V.; Narayanan, S.; Mukund, N.; Arjun, M.; Raunak, K. Development of defect free Monel 400 welds for marine application. *Adv. Mater. Res.* **2012**, *383–390*, 4695. [CrossRef]

9. Ventrella, V.A.; Berretta, J.R.; De-Rossi, W. Micro Welding of Ni-based Alloy Monel 400 Thin Foil by Pulsed Nd:YAG laser. *Phys. Procedia* **2011**, *12*, 347–354. [CrossRef]

10. Kukliński, M.; Bartkowska, A.; Przestacki, D. Microstructure and selected properties of Monel 400 alloy after laser heat treatment and laser boriding using diode laser. *Int. J. Adv. Manuf. Tech.* **2018**, *98*, 3005–3017. [CrossRef]

11. Bopp, C.; Santhanam, K. Corrosion Protection of Monel Alloy Coated with Graphene Quantum Dots Starts with a Surge. *ChemEngineering* **2019**, *3*, 80. [CrossRef]

12. Himanshu, B.S. Surface roughness modeling of Ni -Co based alloy using RSM. *Int. J. Emerg. Trends Res.* **2016**, *1*, 29–39.

13. Davis, J.R. *Nickel, Cobalt, and Their Alloys*, 1st ed.; ASM International: Novelty, OH, USA, 2000; pp. 294–295.

14. Zils, R. Werkstoffe im Pumpenbau. *Chem. Ing. Tech.* **2008**, *80*. [CrossRef]

15. Lee, F.T.; Major, J.F.; Samuel, F.H. Fracture Behaviour Of A112wt.%Si0.35wt.%Mg(O-O.O2)wt.%Sr Casting Alloys Under Fatigue Testing. *Fatigue Fract. Eng. Mater. Struct.* **1995**, *18*, 385–396. [CrossRef]

16. Röhrig, K.E. Guss aus Hochkorrosionsbeständigen Nickel-Basislegirerungen. Available online: https://www.kug.bdguss.de/fileadmin/content/Publikationen-NormenRichtlinien/Guss_aus_hoch_korr. best._Nickelleg.pdf (accessed on 10 March 2020).

17. Seifi, M.; Salem, A.; Beuth, J.; Harrysson, O.; Lewandowski, J.J. Overview of Materials Qualification Needs for Metal Additive Manufacturing. *JOM* **2016**, *68*. [CrossRef]

18. Yap, C.Y.; Chua, C.K.; Dong, Z.L.; Liu, Z.H.; Zhang, D.Q.; Loh, L.E.; Sing, S.L. Review of selective laser melting: Materials and applications. *Appl. Phys. Rev.* **2015**, *2*, 041101. [CrossRef]

19. Marenych, O.; Kostryzhev, A.; Shen, C.; Pan, Z.; Li, H.; Van Duin, S. Precipitation Strengthening in Ni–Cu Alloys Fabricated Using Wire Arc Additive Manufacturing Technology. *Metals* **2019**, *9*, 105. [CrossRef]

20. Labudovic, M.; Hu, D.; Kovacevic, R. A three dimensional model for direct laser metal powder deposition and rapid prototyping. *J. Mater. Sci.* **2003**, *38*, 35–49. [CrossRef]

21. Anderson, R.; Terrell, J.; Schneider, J.; Thompson, S.; Gradl, P. Characteristics of Bi-metallic Interfaces Formed during Direct Energy Deposition Additive Manufacturing Processing. *Metall. Mater. Trans. B* **2019**, *50*, 1921–1930. [CrossRef]

22. Raffeis, I.; Adjei-Kyeremeh, F.; Vroomen, U.; Suwampinij, P.; Ewald, S.; Bührig-Polaczek, A. Investigation of the Lithium-Containing Aluminum Copper Alloy (AA2099) for the Laser Powder Bed Fusion Process [L-PBF]: Effects of Process Parameters on Cracks, Porosity, and Microhardness. *JOM* **2019**, *71*, 1543–1553. [CrossRef]

23. Mugwagwa, L.; Yadroitsev, I.; Matope, S. Effect of Process Parameters on Residual Stresses, Distortions, and Porosity in Selective Laser Melting of Maraging Steel 300. *Metals* **2019**, *9*, 1042. [CrossRef]

24. Mercelis, P.; Kruth, J.P. Residual stresses in selective laser sintering and selective laser melting. *Rapid Prototyp.* **2006**, *12*, 265. [CrossRef]

25. Nguejioa, J.; Szmytkaa, F.; Hallaisb, S.; Tanguyb, A.; Nardonec, S.; Godino Martinezc, M. Comparison of microstructure features and mechanical properties for additive manufactured and wrought nickel alloys 625. *Mater. Sci. Eng. A* **2019**, *764*, 138214. [CrossRef]

26. Pourbabak, S.; Montero-Sistiaga, M.L.; Schryvers, V.; Van Humbeeck, J.; Vanmeensel, K. Microscopic investigation of as built and hot isostatic pressed Hastelloy X processed by Selective Laser Melting. *Mater. Charact.* **2019**, *153*. [CrossRef]

27. Raghavana, S.; Baicheng, Z.; Pei, W.; Chen-Nan, S.; Mui, L.S.N.; Tao, L.; Jun, W. Effect of different heat treatments on the microstructure and mechanical properties in selective laser melted INCONEL 718 alloy. *Mater. Manuf. Process.* **2016**. [CrossRef]

28. Li, C.; White, R.; Fang, X.Y.; Weaver, M.Y.; Guo, B. Microstructure Evolution Characteristics of Inconel 625 Alloy from Selective Laser Melting to Heat Treatment. *Mater. Sci. Eng. A* **2017**. [CrossRef]

29. Kulkarni, A. Additive Manufacturing of Nickel Based Superalloys. *arXiv* **2018**, arXiv:1805.11664.

30. Aminipour, N.; Derakhshandeh-Haghighi, R. The Effect of Weld Metal Composition on Microstructural and Mechanical Properties of Dissimilar Welds between Monel 400 and Inconel 600. *J. Mater. Eng. Perform.* **2019**, *28*, 6111. [CrossRef]

31. Debidutta, M.; Vignesh, M.K.; Ganesh, R.B.; Pruthvi, S.; Devendranath, K.R.; Arivazhagan, N.; Narayanan, S. Mechanical Characterization of Monel 400 and 316 Stainless Steel Weldments. *Procedia Eng.* **2014**, *75*, 24–28.

32. Chlebus, E.; Grubern, K.; Kuźnicka, B.; Kurzac, J.; Kurzynowski, T. Effect of heat treatment on the microstructure and mechanical properties of Inconel 718 processed by selective laser melting. *Mater. Sci. Eng. A* **2015**, *639*, 647–655. [CrossRef]

33. McLouth, T.D.; Witkin, D.B.; Bean, G.E.; Sitzman, S.D.; Adams, P.M.; Lohser, J.R.; Yang, J.-M.; Zaldivar, R.J. Variations in ambient and elevated temperature mechanical behavior of IN718 manufactured by selective laser melting via process parameter control. *Mater. Sci. Eng. A* **2020**, *780*, 139–184. [CrossRef]

34. Marchese, G.; Garmendia, X.C.; Calignano, F.; Lorusso, M.; Biamino, S.; Minetola, P.; Manfredi, D. Characterization and Comparison of Inconel 625 Processed by Selective Laser Melting and Laser Metal Deposition. *Adv. Eng. Mater.* **2017**, *19*. [CrossRef]

35. Yan, F.; Xiong, W.; Faierson, E.J. Grain Structure Control of Additively Manufactured Metallic Materials. *Materials* **2017**, *10*, 1260. [CrossRef]

36. Seede, R.; Mostafa, A.; Brailovski, V.; Jahazi, M.; Medraj, M. Microstructural and Microhardness Evolution from Homogenization and Hot Isostatic Pressing on Selective Laser Melted Inconel 718: Structure, Texture, and Phases. *J. Manuf. Mater. Process.* **2018**, *2*, 30. [CrossRef]

37. Brandt, M.; Sun, S.; Leary, M.; Feih, S.; Elambasseril, J.; Liu, Q. High-Value SLM Aerospace Components: From Design to Manufacture. *Adv. Mater. Res.* **2013**, *147*, 633. [CrossRef]

38. Shifeng, W.; Shuai, L.; Qingsong, W.; Yan, C.; Sheng, Z.; Yusheng, S. Effect of molten pool boundaries on the mechanical properties of selective laser melting parts. *J. Mater. Process. Technol.* **2014**, *214*, 2660–2667. [CrossRef]

39. Dong, Z.; Liu, Y.; Wen, W.; Ge, J.; Liang, J. Effect of Hatch Spacing on Melt Pool and As-built Quality During Selective Laser Melting of Stainless Steel: Modeling and Experimental Approaches. *Materials* **2019**, *12*, 50. [CrossRef] [PubMed]

40. Sun, S.; Brandt, M.; Easton, M. *Powder Bed Fusion Processes: An Overview, Laser Additive Manufacturing-Materials, Design, Technologies, and Applications*; Centre for Additive Manufacturing: Melbourne, Australia, 2017; pp. 55–77.

41. Hanzl, P.; Zetek, M.; Bakša, T.; Kroupa, T. The Influence of Processing Parameters on the Mechanical Properties of SLM. *Parts Procedia Eng.* **2015**, *100*, 1405–1413. [CrossRef]

42. Zhang, B.; Li, Y.; Bai, Q. Defect Formation Mechanisms in Selective Laser Melting: A Review. *Chin. J. Mech. Eng.* **2017**, *30*, 515–527. [CrossRef]

43. Kasperovich, G.; Haubrich, J.; Gussone, J.; Requena, G. Correlation between porosity and processing parameters in TiAl6V4 produced by selective laser melting. *Mater. Des.* **2016**, *105*, 160–170. [CrossRef]

44. Strondl, A.; Lyckfeldt, O.; Brodin, H.; Ackelid, U. Characterization and Control of Powder Properties for Additive Manufacturing. *JOM* **2015**, *67*. [CrossRef]

45. Casati, R.; Lemke, J.; Vedani, M. Microstructure and fracture behavior of 316L austenitic stainless steel produced by selective laser melting. *J. Mater. Sci. Technol.* **2016**, *32*, 738–744. [CrossRef]

46. Carter, L.N.; Martin, C.; Withers, P.J.; Attallah, M.M. The influence of the laser scan strategy on grain structure and cracking behaviour in SLM powder-bed fabricated nickel superalloy. *J. Alloys Compd.* **2014**, *615*, 338–347. [CrossRef]

47. Kunze, K.; Etter, T.; Grässlinc, J.; Shklover, V. Texture, anisotropy in microstructure and mechanical propertiesof IN738LC alloy processed by selective laser melting (SLM). *Mater. Sci. Eng. A* **2014**, *620*, 213–222. [CrossRef]

48. Deev, A.A.; Kuznetcov, P.A.; Petrov, S.N. Anisotropy of Mechanical properties and its correlation with the structure of the stainless steel 316L produced by SLM Method. *Phys. Procedia* **2016**, *83*, 789–796. [CrossRef]

49. Etter, T.; Kunze, K.; Geiger, F.; Meidani, H. Reduction in mechanical anisotropy through high temperature heat treatment of Hastelloy Xprocessed by Selective Laser Melting (SLM). *IOP Conf. Ser. Mater. Sci. Eng.* **2015**, *82*, 012097. [CrossRef]

50. DebRoy, T.; Wei, H.L.; Zuback, J.S.; Mukherjee, T. Additive manufacturing of metallic components–process, structure and properties. *Prog. Mater.* **2018**, *92*, 112–224. [CrossRef]

Review

Microstructure and Mechanical Properties of AISI 316L Produced by Directed Energy Deposition-Based Additive Manufacturing: A Review

Abdollah Saboori *, Alberta Aversa, Giulio Marchese, Sara Biamino, Mariangela Lombardi and Paolo Fino

Department of Applied Science and Technology, Politecnico Di Torino, Corso Duca degli Abruzzi 24, 10129 Torino, Italy; alberta.aversa@polito.it (A.A.); giulio.marchese@polito.it (G.M.); sara.biamino@polito.it (S.B.); Mariangela.lombardi@polito.it (M.L.); paolo.fino@polito.it (P.F.)
* Correspondence: abdollah.saboori@polito.it; Tel.: +39-0110904763

Received: 20 April 2020; Accepted: 6 May 2020; Published: 9 May 2020

Abstract: Directed energy deposition (DED) as a metal additive manufacturing technology can be used to produce or repair complex shape parts in a layer-wise process using powder or wire. Thanks to its advantages in the fabrication of net-shape and functionally graded components, DED could attract significant interest in the production of high-value parts for different engineering applications. Nevertheless, the industrialization of this technology remains challenging, mainly because of the lack of knowledge regarding the microstructure and mechanical characteristics of as-built parts, as well as the trustworthiness/durability of engineering parts produced by the DED process. Hence, this paper reviews the published data about the microstructure and mechanical performance of DED AISI 316L stainless steel. The data show that building conditions play key roles in the determination of the microstructure and mechanical characteristics of the final components produced via DED. Moreover, this review article sheds light on the major advancements and challenges in the production of AISI 316L parts by the DED process. In addition, it is found that in spite of different investigations carried out on the optimization of process parameters, further research efforts into the production of AISI 316L components via DED technology is required.

Keywords: additive manufacturing; directed energy deposition; AISI 316L; microstructure; mechanical properties

---

1. Introduction

In recent decades, additive manufacturing (AM) technologies, also recognized as three dimensional (3D) printing, has attracted significant attention in different industries [1,2]. In principle, in all-metal AM processes, at first, a solid model is sliced in multiple layers to generate a tool path for the printing machine. Thereafter, the 3D component is produced in a layer-wise process according to the sliced model data. In addition to the sliced model, two main elements are required to build a part: a feedstock material (metal powder or wire) and a heat source, which can be a laser, electron beam or electric arc [3]. In general, AM systems are categorized into two different classes: powder bed systems and powder/wire feed systems [4,5]. In powder bed systems, a layer of powder is spread on the building platform or on the previously solidified layer and is selectively fused via an energy source that can be either electron beam or laser beam [6–8]. The ability to produce high-resolution features and internal channels, as well as precision dimensional control, are considered the main advantages of powder bed AM processes [5,9,10]. In contrast, in powder/wire feed systems, the material is fed directly inside a melt pool which is already formed by a focalized heat source on the substrate or on the already deposited layer. Directed energy deposition (DED), as a material feed process, uses a focalized heat source, that can be a laser or an electron beam, and a material which can be powder or wire while being delivered directly into the melt pool. It should also be highlighted that in the literature different names are generally reported for this process [11]. A summary of these names is listed in Table 1.

Table 1. Different commercialized names of the directed energy deposition (DED) process.

Acronym	Technology	Ref.
LENS	Laser Energy Net Shaping	[12,13]
LMD	Laser Metal Deposition	[14,15]
LC	Laser Cladding	[16,17]
DMD	Direct Metal Deposition	[18,19]
LAMP	Laser aided manufacturing process	[3,20]
DLF	Direct Laser Fabrication	[21,22]
LPF	Laser Powder Fusion	[23]

In the DED process the deposition pattern is defined by the relative motion between the substrate and the deposition head. This motion can be obtained by moving only the deposition head, only the substrate, or both substrate and deposition head. The method used mainly depends on the size and the geometry of the substrate [24].

In particular, in the laser-powder DED process, which is known as the most versatile DED process, the powder feeding can be implemented by means of either a single nozzle, coaxial nozzle, or multi-nozzle configuration. The laser-powder DED process has gained considerable interest in recent years thanks to the possibility of repairing parts and production of functionally graded materials (FGM) by varying the alloying element content [25]. In addition to the abovementioned merits of this technology, by using the laser-powder DED process it would be also possible to design specific alloys through the in situ alloying process. In fact, it is possible to deliver various powders into the melt pool simultaneously and, as a consequence, achieve a new composition after solidification. Moreover, a high deposition rate, as well as a rather wide process window, make this process very promising, with respect to the other AM processes, to be employed for the production of large components. These flexibilities of the DED process in the production of net-shape parts and repair of high-value components have broadened its application in various sectors such as aerospace, transportation, medical, and defense.

Despite the aforementioned merits, the DED process has a low powder efficiency and final rough surface that should be machined after the building process. Furthermore, previous works have shown that the thermal history of a part produced via the DED process has a marked effect on the microstructure and mechanical performance of components [26–28]. Therefore, the quality of DED parts is mainly defined by the building parameters used during the process. It is well-known that a large number of parameters can be varied in DED, and these include laser power, scan speed, powder feed rate, building atmosphere, deposition pattern, and many others.

In recent years, stainless steels have been intensively processed by AM technologies, mainly owing to the high mechanical properties that make them suitable for a wide range of applications in various industries such as the automotive, aerospace and petrochemical sectors [4,29,30]. AISI 316L steel is by far the most processed and studied, and its success is mainly related to its weldability, corrosion resistance, and tensile properties. Among the AM technologies, the DED process, which can provide a high grade of flexibility in the design and production of large AISI 316L components, could attract significant attention. In fact, large complex parts can be produced via the DED process with a reduction in the weight, the waste of expensive starting material, and the number of costly post-machining steps.

In recent years, a growing body of literature has emerged in which the correlation among process parameters, microstructure, and mechanical properties of DED AISI 316L stainless steel (Figure 1) has been studied [31–35]. For instance, Yadollahi et al. studied the influence of the time interval between the deposition of layers and the mechanical properties of DED AISI 316L [36]. Saboori et al. investigated the effect of powder recycling and deposition pattern on the microstructure and mechanical properties of AISI 316L samples produced by the DED process [37,38]. Zheng et al. evaluated the effect of DED process parameters on the evolution of the dimensional and surface quality, microstructure, internal surface, and mechanical performance of AISI 316L samples [39]. Terrassa et al. studied the role of hatch rotation angle on the built quality of DED AISI 316L samples [40]. Tan et al. examined the correlation between the porosity, density, and microstructure of AISI 316L samples produced via DED technology [41]. However, it should be highlighted that according to the existing body of literature on the processing of AISI 316L stainless steel via the DED process, microstructure and mechanical properties analyses are the dominant features that have been considered in the previous research (Figure 2). Hence, this paper provides an overview of the microstructure and properties of DED AISI 316L, and summarizes the main effects of the building parameters on the quality of the final products. In fact, the aim of this article is to review the additive manufacturing of AISI 316L alloy by DED in terms of microstructural development and mechanical properties of the samples produced with the optimal process parameters. First, the role of various factors, such as thermal history and process parameters, on the microstructure of DED AISI 316L is reviewed, and thereafter, the influence of different parameters, such as building direction, building parameters, and powder quality, on the mechanical properties of manufactured components is discussed. In general, the target here is not to assemble all of the existing literature about DED of AISI 316L alloy, but to clarify the importance and opportunities of this innovative process in this field.

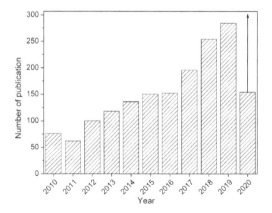

Figure 1. Publication trend in DED AISI 316L chronology between 2010 and 2020 in ScienceDirect (This graph is plotted according to a simple search of "316L" + "DED" at www.sciencedirect.com).

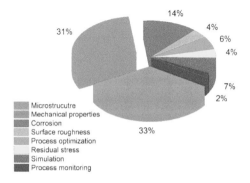

Figure 2. Pie chart showing the percentage of publications at www.sciencedirect.com on different features of the processing of AISI 316L via the DED process.

2. Microstructure

As a thermal process, DED includes a sequence of physical phenomena, such as rapid heating, melting, potential vaporization, and rapid cooling [39]. However, to date, the stability of the microstructure of components produced via DED, which takes place under non-equilibrium conditions, is poorly understood.

In general, the thermal history of DED components, such as the high heating/cooling rate, marked temperature gradient, and bulk temperature increment, define the morphology and grain size of DED components. However, since all of the process parameters and variables have a significant influence on the thermal history of parts, the prediction of microstructural characteristics and their dependence degree remains a significant challenge for metallic materials processed by DED. Nevertheless, to have an effective control mechanism to produce metallic components via DED with excellent mechanical characteristics, it is necessary to overcome this challenge. Therefore, in the literature, several authors have studied the role of specific parameters on the microstructural features and mechanical properties of metallic components produced via DED [4,42–44].

Local solidification rates, the temperature gradient at the liquid/solid interface (G), and the ratio of cooling rate/thermal gradient (R) are the effective parameters that define the final solidified microstructure. In fact, G/R and G × R are found to be the most critical solidification parameters that have a marked influence on the shape of the liquid/solid interface and on the size of microstructure, respectively [45,46]. Substantially, after solidification of metallic parts produced by DED, columnar grains, which represent an elongated morphology that grows in the direction of a maximum thermal gradient, columnar-equiaxed grains, and equiaxed grains are the three structure morphologies that can be formed as a consequence of various G and R values [23]. For instance, it has been revealed that higher solidification rates promote a transition from columnar gain morphology to an equiaxed morphology, and an increment in the cooling rate results in microstructure refinement [27]. Moreover, it should be noted that G/R plays a vital role in the final microstructural morphology, in which low G/R values result in equiaxed structures and high G/R values in columnar structures (Figure 3) [45]. In general, it is found that with cooling rates of 10^3 to 10^4 K/s, it would be possible to achieve desired microstructure and mechanical properties in components produced via DED [23,47]. Part geometry, environmental conditions, and material characteristics are the main factors that can have a marked influence on the optimal G and R values [45].

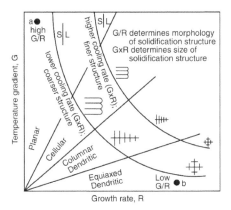

Figure 3. Solidification map showing the effect of temperature gradient and growth rate on the morphology and size of the resulting microstructure [48].

To date, several research studies have been carried out to determine the most effective type of heat transfer mechanism and, consequently, the cooling rate within the AISI 316L components produced by

DED [37,49]. As an example, Saboori et al. showed that in an AISI 316L sample different heat transfer mechanisms dominate in different zones of a melt pool with the formation of various microstructural features [37]. In the central part of the melt pool, where the liquid metal solidifies slightly later, the convective heat transfer mode dominates, whereas at the melt pool borders and across the heat affected zone (HAZ), the solid conduction heat transfer mode is the effective heat transfer mode. In addition, it is reported that at the edges of the laser track, where the lateral sides of the melt pool are exposed to the environment, a complex mix of convective–conductive–radiating heat transfer occurs. Figure 4 reports the general microscopic images of the representative microstructures of as-built AISI 316L samples produced via DED. The first visible feature is the curved border of melt pools, which is the typical AM microstructural characteristic as a consequence of the Gaussian distribution of laser energy (Figure 4a). It is also clear that the temperature gradients in the direction perpendicular to the curved melt pool borders are intense and accordingly lead to the formation of a marked directional growth of the dendrites from the melt pool borders and converging towards the center of the melt pool (Figure 4b). On the contrary, at the central part of the melt pool, owing to the change of heat transfer mode, equiaxed dendrites are more likely to form. Regarding the whole section of the deposited component, as a consequence of the complex heat transfer during the DED process of this alloy, it is found that the columnar structure growing in the direction of the maximum thermal gradient dominates in the middle height of the sample, whereas in the last deposited layers the cellular structure is present [37]. This variation in the microstructure of components along the building direction results in the oscillation in the microhardness values along the building direction. The microhardness of the material decreases at the beginning from the first deposited layer to the second, and thereafter increases gradually toward the last layers [50,51]. This variation in the microhardness of components is found to result from the different velocity of solidification in the sample. In addition, during the deposition, owing to the reheating of previously deposited layers, the middle area is also exposed to cycle reheating that results in the formation of a HAZ area which remains at higher temperatures for a longer period of time. Thus, finer microstructure and higher microhardness are expected for the bottom and top of the DED components which undergo higher cooling rates with respect to other areas. However, the research found a negligible porosity in the final microstructure, even after the optimization of the process parameters.

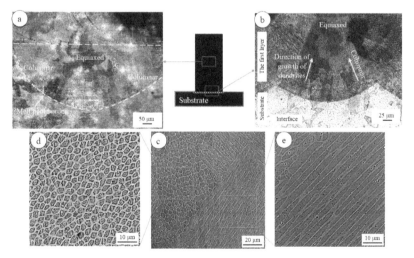

Figure 4. Light optical microscopy micrographs of the DED AISI 316L steel samples produced by DED: (**a**) a representative melt pool at the middle height, (**b**) the microstructure of the first layer, (**c**) SEM images of the columnar and equiaxed microstructures referring to the last deposited layer, (**d**,**e**) high magnifications of (**c**) from two different regions [37].

In another work, Bi et al. studied the microstructure of AISI 316L thin walls produced via DED (Figure 5) [52]. In addition to the typical microstructural characteristics that can be formed in the DED samples, they also reported that the remelting and tempering of the middle layers during the deposition of the next layer, with the exception of the previous layer, are the source of microstructural variations [52].

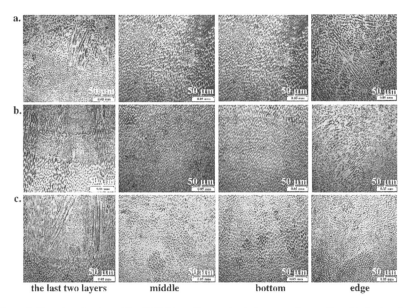

Figure 5. The microstructure of the DED thin walls examined at cross-section and longitudinal section in the middle of the thin wall. (**a**) With a constant laser power of 300 W, (**b**) Process control with a constant set-value 0.5 V and (**c**) Process control with a path-dependent set-value 0.3 V (2 mm)–0.5 V (56 mm)–0.3 V (2 mm) [52].

Kruz [53], Kelly and Kampe [54], and Colaco and Vilar [55] suggested that the microstructural features and mechanical characteristics of DED parts depend partially on the solid-state transformation during the cooling step. However, it is reported that these transformations are driven by the thermal cycles that the material undergoes during the deposition. Since one of the most critical parameters that can affect the thermal history, and accordingly the microstructural evolution of metallic materials, is the cooling rate, several studies have been undertaken to estimate this parameter during DED of metallic materials. For this reason, several experiments [56,57], including analytical and numerical [58,59] approaches, have been developed to predict the effects of process variables on the cooling rate and consequently on the resulting microstructure in DED samples. For instance, Hofmeister et al. analyzed the thermal gradient and cooling rates in the regions near the melt pool through the monitoring of the melt pool via a digital video camera with thermal imaging techniques [60]. Griffith et al. evaluated the in situ thermal history of DED samples by inserting a thermocouple directly into the sample [42]. All of the experimental results showed that the formation of a very fine microstructure in DED components is a direct consequence of high cooling rates and the temperature gradient [56,57]. Gosh and Choi found that since the cooling rate significantly influences the primary cellular arm spacing (PCAS), it is possible to evaluate the thermal history and cooling rate of DED parts via PCAS analysis [61]. Therefore, they proposed an equation which describes the correlation between the PCAS and the thermal history of DED samples. Subsequently, several research studies have been conducted to investigate the correlation between the cooling rate (ε) and PCAS (λ), and outcomes reveal a linear relationship between logλ and logε [62,63]. Bontha et al. also studied the correlation between dendrite morphology, temperature gradient, and solidification rate during the DED process [27]. Recently, Saboori et al. evaluated the PCAS of AISI 316L alloy produced via DED at different distances from the substrate [37]. Thereafter, in their work, the cooling rate is estimated as a function of sample height using the PCAS values. It is found that by increasing the height of the sample up to the last layers, the PCAS values monotonously increase, and thereafter values drop suddenly when close to the previous layers (Figure 6).

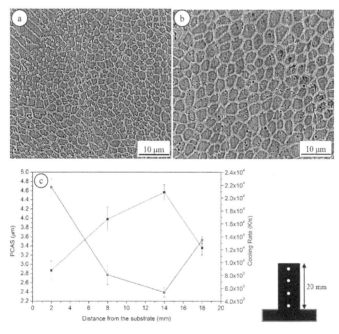

Figure 6. SEM images of DED AISI 316L at: (**a**) 2 mm, (**b**) 14 mm from the deposition/substrate interface, (**c**) primary cellular arm spacing (PCAS) and cooling rate as a function of the distance from the substrate [37].

As can be seen in Figure 6, the estimated cooling rate during DED of AISI 316L lies in the range of 10^3–10^4 K/s, which is in agreement with the typical cooling rate reported for the DED process [37,49,51]. Ma et al. compared the thermal history, cooling rate, and microstructural features of AISI 316L stainless steel produced by DED and laser powder bed fusion (LPBF) [49]. Indeed, in their work, five sets of process parameters are used to produce different cubes. Figure 7 shows the 3D view of the microstructures of the five different cubes; (a) low-power LPBF, (b) high-power LPBF, (c) small-size DED, (d) middle-size DED, and (e) large-size DED (Figure 7). As can be seen, the microstructure of all the samples is almost fully composed of cells, and all the dendrites have transformed into cells. Thereafter, their analyses revealed that the PCAS of each sample gradually increases with increasing the energy density (e) (Figure 8). Ma et al. also found that the solidification behavior of samples produced via DED is slightly different from that of specimens processed by LPBF. This means that by increasing the cooling rate, the solidification behavior is far from representing equilibrium conditions and accordingly results in a non-equilibrium microstructure. Moreover, it is revealed that lower cooling rate and lower supercooling in DED result in the formation of grains with a width and length 2–5 times coarser than those formed in the LPBF process.

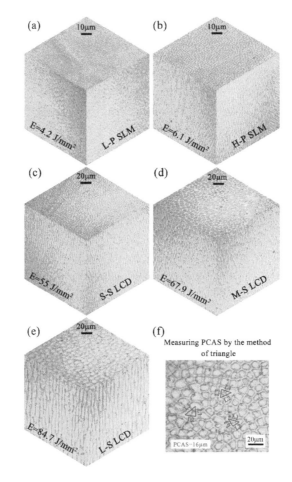

Figure 7. 3D composite view showing the cellular morphologies of the five kinds of typical samples at the processing parameters of (**a**) low-power laser powder bed fusion (LPBF), (**b**) high-power LPBF, (**c**) small-size DED, (**d**) middle-size DED, and (**e**) large-size DED. (**f**) Schematic sketch of measuring PCAS by the three-angle method [49].

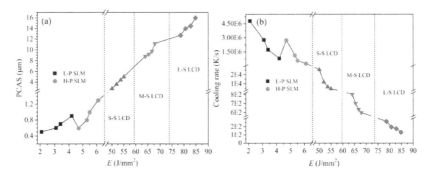

Figure 8. The effect of energy density E on the (**a**) PCAS and the (**b**) cooling rate of the as-forming AISI 316 L stainless steel samples at different processing parameters by LPBF and DED [49].

All of the experiments conducted in different studies proved that the size of dendrite arms would be in the range of a few microns. For example, Hofmeister et al. found that the average PCAS of AISI 316L produced via DED increased from 3 to 9 µm when the laser power increases [42]. Table 2 compares all the dendrite sizes that have been found in different studies. It should be highlighted that the evaluation of PCAS has been carried out using SEM images and three-angle method [37,49,64].

Table 2. A summary of PCAS reported for AISI 316L alloy processed by DED.

Author	Dimension (µm)	Ref.
Saboori et al.	2.8–4.8	[37]
Song et al.	1.3–3.0	[65]
Hofmeister et al.	3.25–8.68	[42]
Syed et al.	<5	[66]
Zheng et al.	8–20	[51]
Smugeresky et al.	2–15	[67]

In addition to the effect of cooling rate on microstructural morphology, it should be noted that this rapid solidification process also results in the phase composition of the microstructure of the samples in the as-built condition. In general, in the standard rapid solidified austenitic stainless steel, two distinct microstructural constituents can be achieved: austenite (γ) and δ-ferrite [68]. However, in order to predict the microstructure of the material from the phase composition point of view, some chemical composition-based phase diagrams, such as Schäffler and DeLong diagrams, are employed. The Schäffler diagram is the most accepted diagram used to predict the final microstructure of the material. In fact, this diagram is commonly employed to estimate the δ-ferrite content in the final microstructure of AISI 316L (Figure 9a). In addition to predicting δ-ferrite content, this diagram is also capable of predicting the existence of ferrite, martensite, and austenite phases in AISI 316L alloy as a function of the Cr and Ni equivalents [69]. However, it is reported that this diagram is not an exact diagram and, thus, the outcome is an approximation of the final δ-ferrite content [69]. Zhi'En et al. also studied the solidification mode of the AISI 316L alloy according to the role of alloying elements [41]. In their work, it is shown that by changing the ferrite stabilizer or austenite stabilizer content the solidification mode is different and, consequently, the final alloy can be either duplex or fully austenitic (Figure 9b).

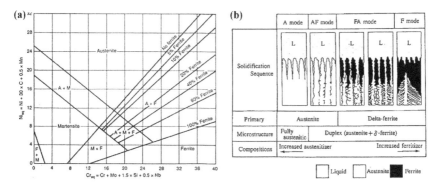

Figure 9. (**a**) Schaeffler's diagram and (**b**) austenitic steel solidification diagram [41].

However, it should be noted that this diagram has typically been used to predict the microstructure of corrosion-resistant steels with carbon content up to 0.25% after the welding process. Since the cooling rate is always higher than the welding process in the DED process, the solidification is not under equilibrium conditions. Therefore, it is highly recommended that, in addition to the conventional

Schäffler diagram, another diagram, namely the pseudo-binary predictive phase diagram, is used [70]. This means that once the theoretical δ-ferrite content is defined by Schäffler diagram, the WRC-1992 modified Cr and Ni equivalent formulas, which are shown on the X- and Y-axes of Figure 10a, can be used to place the steel being investigated in the pseudo-binary phase diagram (Figure 10).

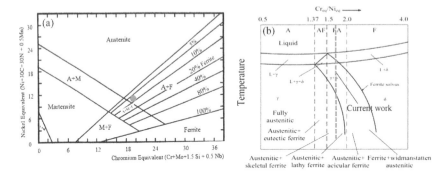

Figure 10. (**a**) Schaeffler constitution diagram showing the location of the composition of the AISI 316L stainless steel powders, (**b**) pseudo-binary phase diagram [37].

For instance, Tan et al. studied the microstructure of AISI 316L produced by DED. At first, they found that according to the chemical composition of the material and the Schäffler diagram, the estimated δ-ferrite content of the columnar boundary lies in the region of 20% δ-ferrite. The Cr_{eq} and Ni_{eq} of their DED AISI 316L suggest that the solidification of their material falls within the upper range of the austenitic-ferritic mode in such a way that an austenitic steel forms a predominantly austenitic microstructure with columnar δ-ferrite along the solidification direction (Figure 11) [41]. The δ-ferrite phase can be recognized by the higher Cr and Mo content. Milton et al. studied the microstructure of DED AISI 316L and revealed that Cr and Ni equivalent contents lie in the range of 5–10% δ-ferrite [71]. Zietala et al. also investigated DED of AISI 316L and found that intercellular δ-ferrite formed at sub-grain boundaries as a consequence of its enrichment of Cr and Mo, and depletion of Ni [35].

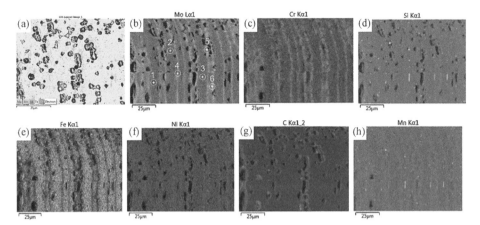

Figure 11. Element mapping for deposition body area: (**a**) layered element image, (**b**) molybdenum and target body, boundary element composition points, (**c**) chromium, (**d**) silicon, (**e**) iron, (**f**) nickel, (**g**) carbon, (**h**) manganese [41].

Formation of a duplex microstructure during DED of 316L was also revealed by Saboori et al. [37]. Indeed, they showed that the interdendritic arms are enriched in molybdenum and chromium, which are δ-stabilizer elements. This elemental distribution can also be explained by the cooling rate of the process in which, during rapid solidification, austenite-promoting elements such as Ni and C are consumed to solidify the austenite phase and then the residual liquid phase is enriched in δ-stabilizer elements in the interdendritic regions.

Inclusion formation, such as of oxides rich in Si or Si and Mn, is an undesirable feature reported in DED of 316L [72]. These oxide structures are typically found during the ladle practice of high-content Mn/Si steels [73,74]. It is found that, due to their very high reactivity with oxygen, the formation of these oxides, even in secondary steelmaking, is difficult to avoid. Thus, in spite of using protective shielding gas to protect the melt pool, finding these kinds of oxides is not surprising. However, it should be noted that their detrimental effect on the components produced via DED with respect to the conventional steelmaking processes is relatively low, mainly owing to their final reduced size and spherical shape [37]. Lou et al. studied the oxide inclusion in laser AM of AISI 316L and reported their detrimental effect on the toughness and stress corrosion cracking behavior [75]. In fact, in their work, intergranular Si-rich and Si/Mn-rich oxides were found in as-built AISI 316L components (Figure 12). In another study, Ganesh et al. investigated the corrosion behavior of AISI 316L produced via DED [76]. They found that the presence of these oxides in the final microstructure weakens the corrosion resistance of the AISI 316L parts.

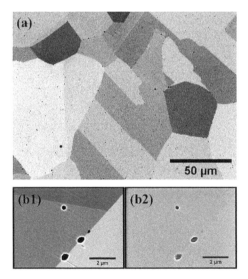

Figure 12. Oxide inclusions in the fully-recrystallized additive manufacturing (AM) AISI 316L stainless steel: (**a**) low magnification back-scattered electron image; (**b1**) high magnification back-scattered electron image; (**b2**) high magnification secondary electron image [75].

The size and the composition of the oxides found in AM AISI 316L samples, together with their effect on the material properties, are reported in Table 3. All these findings prove that in order to achieve the full potential of AISI 316L alloy in different applications, a more reliable deposition atmosphere should be used to comprehensively protect the melt pool and consequently reduce the oxide content of the alloy.

Table 3. A summary of the composition and size of oxides found in AM AISI 316L alloy.

AM Technology	Composition	Size	Effect	Ref.
LPBF	Si/Mn and Si/Mn/Mo rich oxides	50 nm–1 mm	Detrimental effect on toughness and stress corrosion cracking	[75]
DED	Cr_2O_3, MnO and SiO_2	0.31–0.49 μm	Higher yield strength	[77]
DED	Mn/Si-rich oxides	-	Detrimental effect on the elongation	[37]
DED	MnO and SiO_2	-	Possible effect on ductility reduction	[78]

However, it should be highlighted that since the size and quantity of these inclusions are normally lower than the resolution of the X-ray diffraction (XRD) analysis, they have mainly been analyzed via SEM and image analysis. The formation of these inclusions, mainly oxides, can change the failure behavior of material from ductile mode to brittle mode (as for the composite materials) [79,80].

3. Mechanical Properties

Mechanical properties can be considered one of the main indicators of the quality of an AM process. Hardness and tensile properties are, in fact, often used as key performance indicators (KPI) of AM components. From an industrial point of view, indeed, tensile samples are generally built together with AM components in order to validate the building process. Because of these reasons, mechanical properties of AM materials have been deeply investigated, and several tensile and hardness analyses of as-built DED AISI 316L samples obtained using different building or post-processing conditions are available in the literature.

Table 4 reports mechanical properties (i.e., Vickers hardness, Hv; yield strength, YS; ultimate strength, US; elongation, (ε)) of DED AISI 316L steel samples produced in different conditions (i.e., with several power (P) and scan speed (V) values, built along a perpendicular (V) or parallel (H) direction with respect to the building platform, with new powder or re-used powder, with Ar or N_2 shielding gas (SG) or a build chamber (BC) as protective atmosphere). These are compared with the tensile properties of AISI 316L samples obtained by conventional technologies (CT).

Table 4. A summary of the mechanical properties of AISI 316L samples produced via different technologies.

	P (W)	V (mm/s)	Direction	Gas	Hv	YS (MPa)	US (MPa)	ε (%)	H_c	Ref.
CT			Hot rolled		-	360	625	69	0.74	[81]
			Cast		170	310	620	45	1.00	[82]
Building parameters	1600	28	-	Ar BC	250	430	650	43	0.51	[49]
	3400	10	-	Ar BC	210	370	590	36	0.59	[49]
	4600	5	-	Ar BC	190	300	560	31	0.87	[49]
	*	2	H	Ar SG	310	505	625	19	0.24	[83]
	*	10	H	Ar SG	370	610	690	24	0.13	[83]
	600	*	H	Ar SG	350	585	655	18	0.12	[83]
	1400	*	H	Ar SG	320	545	620	19	0.14	[83]
Building direction	2000	8.3	V	-	-	415	770	6.5	0.86	[70]
	2000	8.3	H	-	-	580	900	4	0.55	[70]
	-	-	V	Ar SG	-	352	536	46	0.52	[83]
	-	-	H	Ar SG	-	558	639	21	0.15	[83]
	400	15	V	-	272	479	703	46	0.47	[35]
	400	15	H	-	289	576	776	33	0.35	[35]
	360	16	V	Ar BC	220–260	538–552	690–703	35–38	0.28–0.27	[84]
	360	16	H	Ar BC	220–260	448–455	545–634	4–25	0.22–0.39	[84]
Powder quality	-	-	H **	N_2 SG	-	469	628	31	0.34	[37]
	-	-	H ***	N_2 SG	-	458	652	16	0.42	[37]

Table 4. *Cont.*

	P (W)	V (mm/s)	Direction	Gas	Hv	YS (MPa)	US (MPa)	ε (%)	H_c	Ref.
	328	17	V	Ar BC	-	-	550	-	-	[39]
	360	16	V	Ar BC	222–260	448–455	545–634	4–25	0.22–0.39	[84]
Atmosphere	400	15	V	Ar SG	-	352	536	46	-	[83]
	-	-	H	N_2 SG	-	469 ± 3	628 ± 7	31 ± 2	0.34	[64]
	-	-	H	N_2 BC	-	530 ± 5	670 ± 6	34 ± 1	0.26	[64]

* The other building parameters were selected based on an orthogonal experimental design, the mechanical properties are intended to be the results of the application of the statistical method, ** Fresh powder, *** Re-used powder.

From the comparison of the data, it is evident that the tensile strength of DED AISI 316L samples is generally higher than that of conventionally manufactured steels. The reason for these peculiar mechanical properties of as-built AISI 316L parts can be found in their unique microstructure. The main factors that allow the achievements of high mechanical properties are the reduced grains and dendrite size, the presence of residual δ ferrite, and the presence of a dense dislocation network. These factors can also explain the low ductility values of the deposited parts. As previously discussed, these microstructural features are strictly correlated to the high cooling rate which the material undergoes while being processed.

The strengthening effect of the refined microstructure can be correlated to the well-known Hall–Petch equation that associates the material grain size and the YS as follows:

$$YS = YS_0 + \frac{k}{\sqrt{d}} \tag{1}$$

where YS_0 is the frictional stress resisting the motion of gliding dislocations in the absence of grain boundaries, d is the grain size, and k is a material constant.

Yan et al. suggested that for AM AISI 316L parts, there is a Hall–Petch type strengthening effect that correlates the YS to the cell size rather than to the grain size [73].

Yadollahi et al. claimed that metastable δ-ferrite also plays a vital role in the determination of the mechanical properties of DED AISI 316L samples as it is harder than the austenitic matrix [78]. Guo et al. also attributed the higher YS and US values obtained in their work to the presence of the high-temperature δ-ferrite phase. The ferritic phase causes a refinement of the microstructure and a consequent reduction of crack propagation [37]. The higher tensile properties associated with the presence of δ-ferrite are also correlated to the internal strain hardening that arises during the AM process caused by the different coefficient of thermal expansion of the two phases.

The high mechanical properties of AM AISI 316L were also ascribed to the high dislocation density of the as-built material [85]. Saeidi et al. observed a reduction in hardness of LPBF AISI 316L as a consequence of an annealing heat treatment. The microstructural analyses did not reveal any change in the microstructure but only a substantial reduction of the dislocation density at the cell boundaries and surrounding the non-metallic inclusions. The dislocation networks were then considered responsible for the high mechanical properties of as-built AM parts [85].

Several authors also highlighted that the plastic region of DED AISI 316L samples is relatively flat, indicating that these samples have a low strain hardening ability with respect to conventionally processed samples [39,84,86]. The strain hardening ability (H_C) is generally calculated as [86]:

$$H_c = \frac{(US - YS)}{YS} \tag{2}$$

By comparing the tensile data of DED AISI 316L steel samples reported in Table 4, it can be noted that there is a wide variation in their mechanical properties. Vickers hardness values, for example, vary between 190 and 370 HV, while the US values fall in the range 536–900 MPa. Nonetheless, the strongest variations can be found in the YS and the ε values, which vary from 300 to 610 MPa and from 4% to 46%, respectively. These discrepancies are generally due to the AM building conditions, powder quality, and tensile sample direction and geometry. Consequently, most of the considered studies related to the mechanical properties of these materials were focused on the understanding of the effect of the building conditions on the YS, US, and ε values. The main investigated aspects are described and summarized here.

3.1. Building Parameters

It is well-known that the DED building parameters, such as laser power, scan speed, and powder feeding rate are key factors that determine the quality of components. The identification of the most suitable combination of these process parameters can assure not only the achievement of dense samples but also can result in the formation of the desired microstructure and mechanical properties [31,87]. As a result, many authors have investigated the effect of combinations of process parameters on mechanical properties of components. In the first rows of Table 4, data related to samples built with different parameters are reported according to the paper of Ma et al. Considering that other DED parameters and machine configurations, such as laser spot size and standoff distance, can also affect the mechanical properties, data recorded in different research studies are not compared with each other. In this way it is possible to underline the influence of process parameters on mechanical behavior of DED parts without neglecting some factors specific to the system employed in each study.

As demonstrated by Ma et al., there is a clear correlation between the tensile properties of DED AISI 316L steel samples and the delivered energy density. The lowest YS, US, and ε (300 MPa, 560 MPa, and 31%, respectively) values were in fact obtained with the highest power and the lowest scan speed (Table 4). The authors explained this effect through the correlation that exists between YS, PCAS, and the width and length of the columnar grains. As previously described, these microstructural features are strictly connected to the cooling rate and thermal gradient, generally controlled trough the building parameters [49].

Similar results were obtained by Zhang et al., who studied the effect of the laser power and scan speed on the hardness values and tensile properties of DED AISI 316L samples [83]. It was demonstrated that in all the samples, the higher the laser power, the lower are YS and US, while the opposite is true for the scanning speed (Table 4). This effect was also explained by the different cooling rate of the materials under different building conditions. High power and low scan speed lead, in fact, to the formation of larger melt pools and lower cooling rate. Low power and high scan speed, on the contrary, lead to the formation of smaller melt pools that solidify with an extremely high cooling rate. The authors did not report, however, any clear correlation between energy input and elongation.

3.2. Building Direction

The correlation between the building direction and tensile properties is a crucial aspect in the determination of the mechanical properties of AM samples. The anisotropy of the microstructure and the presence of some defects affect the stress–strain curve [49]. The main factors related to the building direction that have an effect on the tensile properties can be listed as follows:

- Grain morphology
- Texture
- Elongated dendrites
- Lack of fusion defects

Guo et al. investigated the impact of the building direction of the mechanical properties and showed that the highest mechanical properties (YS, US, and ε) were observed in the H samples (Table 4). The authors attributed this difference to the superior metallurgical bonding of these samples along the tensile direction. Horizontal samples have at least one layer in which the deposition direction is parallel to the tensile direction. The consolidated scan tracks deposited along the direction parallel to the tensile direction act, therefore, as fibers that reinforce the materials during the tensile tests [70].

On the contrary, the V samples might contain some critical elongated defects perpendicular to the tensile direction due to a lack of fusion between different layers, which can be detrimental during the tensile test.

Similar results were obtained by Ziętala et al. and Zhang et al., who confirmed that the highest YS and US were achieved in the H direction. These authors, however, observed higher elongation values in the V samples (Table 4) [35,83]. The higher elongation of vertical samples was attributed by Zhang et al. to the improved ductility due to the dendrites along the growth direction [83]. Yang et al. also analyzed the tensile properties of DED AISI 316L samples (both vertical and horizontal) extracted from a more complex geometry and also obtained lower YS and US in the vertical samples [84]. It is important to underline that these data showed a large scattering, probably because of the different distribution of defects due to the complex geometry from which samples were extracted.

In this case, the vertical samples had a drastically lower elongation value, which was attributed to the delamination phenomenon that was clearly detectable by the fracture surface analyses (Figure 13). The fracture surface analysis indicates that the V and H samples showed different deformation characteristics. The vertical samples had a very reduced necking and displayed a fracture surface characterized by some features, such as unmelted particles and smooth areas, that indicate an incomplete fusion during the building process. The horizontal samples, on the contrary, have a strong necking phenomenon and display the typical ductile fracture surface characterized by fine dimples with a size comparable to the PCAS.

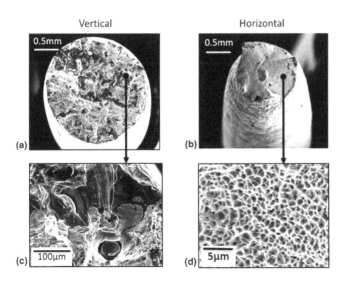

Figure 13. SEM micrographs of the tensile fracture surface: (**a**,**c**) vertical; (**b**,**d**) horizontal [84].

In general, based on these results, it can be concluded that the highest YS and US are generally achieved in the horizontally built samples, and the effect of building direction on the elongation value is still controversial. This discrepancy can be probably related to the other building conditions, as well as the porosity and inclusion content, together with the effect of other strengthening phenomena.

In the case of horizontal samples, however, further studies are needed to investigate the effect of the distance from the substrate on the tensile properties. Wang et al. demonstrated that YS and US of DED 304L stainless steel increase as the distance from the substrate decreases due to the different microstructures that solidify based on the cooling rate [31].

3.3. Powder Quality

It is well known that powder quality plays a fundamental role in the quality of the AM processes as it influences not only consolidation phenomena and consequently porosity contents, but also microstructure and alloy composition.

Saboori et al. studied the effect of powder recycling on the quality of DED AISI 316L parts [37]. In their comparison, the authors found that the mechanical properties of DED AISI 316L samples built with fresh and recycled powders were different (Table 4). The main difference was found in the elongation values that resulted in being very low for the samples built with the recycled powder. This reduction in ε was mainly attributed to the presence of large non-metallic inclusions, which were found to be Mn and Si-based oxides (Figure 14a,b). As demonstrated from the comparison with Figure 14c,d, the inclusions were observed on the recycled particles, suggesting that the oxidation might arise on the particles that are partially heated by the laser beam and exit from the protected atmosphere generated by the shielding gas.

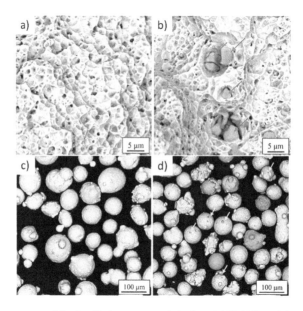

Figure 14. A summary of the tensile fracture morphologies of AISI 316L stainless steel samples produced by DED using (**a**) fresh powder and (**b**) recycled powder. SEM images of (**c**) fresh and (**d**) recycled AISI 316L powders [37].

This marked effect of the oxide content and powder quality on the final mechanical properties of the DED AISI 316L samples suggests that the production of samples in a protective chamber may enhance the quality of the recycled powder, with the deposition allowing the production of parts with higher mechanical properties.

3.4. Building Atmosphere

A close look at Table 4 highlights that DED AISI 316L samples can be built in different atmospheres, such as argon or nitrogen. Furthermore, the protective atmosphere can be generated in two main ways. The first is the use of a simple shielding gas that locally protects the molten material from oxidation, while the second involves the usage of a protective chamber filled with an inert gas.

Aversa et al. recently compared the mechanical properties of DED AISI 316L built using a N_2 shielding gas or using a N_2-filled build chamber. The results showed that very high mechanical properties can be obtained in both conditions and that the use of the Glove Box (GB) allows the achievement of higher YS, US, and ε (Table 4). This effect was attributed to the reduced size of the oxides and the higher N content of BC samples, which have a strengthening effect [64].

The inert gas composition may also have an effect on both the processability of the alloy and its microstructure and mechanical properties. These effects have not yet been investigated in the AM field but were recently studied and reported in the welding literature [88]. In the case of AISI 316L production by DED, it is important to consider that nitrogen is a γ-stabilizer and therefore reduces the high-temperature δ-ferrite phase content which, as previously stated, plays a vital role in the determination of mechanical properties of the final parts. Furthermore, N is an important alloying element for austenitic stainless steel and using it as protective gas might increase the mechanical properties of DED AISI 316L [89].

3.5. Heat Treatment

It is well-known that as-built AM samples and components are characterized by the presence of high residual stresses due to the complex thermal history to which the material is subjected while being processed. Because of this reason, AM components usually undergo specific post-heat treatments that allow the reduction of internal stresses and the homogenization of microstructures.

However, to date, only a small number of studies have been carried out on the effect of the stress-relieving/annealing heat treatments on DED AISI 316L properties [78,90]. The main tensile tests results are summarized in Table 5. The data show that, typically, the YS and US of DED AISI 316L parts are reduced as a consequence of heat treatments. This reduction was mainly attributed to the decrease in the δ-ferrite content [78] and to the reduction of the dislocation density [85]. Furthermore, it is also interesting to underline that a higher strain hardening was observed in heat-treated samples; this can also be due to different dislocation contents.

Table 5. Tensile properties of DED AISI 316L samples in the as-built and heat-treated conditions.

P (W)	V (mm/s)	Conditions	YS (MPa)	US (MPa)	ε (%)	H_c	Ref.
360	8.5	As-built	405–415	620–660	32–40	0.49–0.63	[78]
		1150 °C 2 h Air quenched	325–355	600–620	42–43	0.69–0.91	
380	-	As-built	-	720	56		[90]
		1060 °C 1 h Vacuum treated		605	78		

Despite the differences in the mechanical properties, the fracture surfaces of as-built and heat-treated samples are usually very similar (Figure 15). Morrow et al. reported, for example, a ductile fracture surface characterized by the presence of micrometric dimples for both as-built and heat-treated samples [90]. Moreover, in both samples, extremely fine Mn/Si oxide particles can be detected in the dimples.

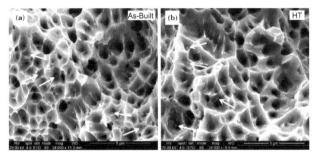

Figure 15. Fracture surfaces of AISI 316L stainless steel: (**a**) as-built and (**b**) heat-treated. Arrows mark a few of the many examples of fine particles observed resting inside ductile dimples throughout the fracture surface [90].

Other more specific mechanical properties of DED 316L were also recently investigated by several authors. Xue et al. and Ganesh et al. studied, for example, the DED 316L impact and fatigue performance of as-built parts [91,92]. Their main finding is that DED samples have a Charpy impact energy and fatigue crack growth rate similar to conventionally manufactured samples. Furthermore, crack propagation in DED samples is transgranular and accompanied by a strain-induced martensite formation.

4. Conclusions

As one of the most employed AM technologies, DED offers excellent potential for the production of complex shape components, which are arduous to produce through conventional processes. AISI 316L is a well-known austenitic stainless with high corrosion resistance, as well as good mechanical properties, which make this alloy an excellent candidate for several sectors, such as the automotive and petrochemical industries. This review article summarizes the latest research carried out to evaluate microstructures and mechanical properties of AISI 316L stainless steel processed by DED. The correlation between the DED process parameters, thermal history, and microstructure of AISI 316L materials is discussed in detail. It is found that most previous works have aimed to determine the optimal process parameters for the DED production of AISI 316L components. These efforts have been taken in order to improve the density and mechanical properties of AISI 316L components produced via the DED process through the control of their microstructure. However, it should be highlighted that, in spite of this research effort, a number of challenges remain that should be considered and addressed in further investigations. The main challenges are associated with the correlation among DED process parameters, thermal history, microstructure, and mechanical characteristics of the DED AISI 316L parts. To date, investigations of DED AISI 316L materials have demonstrated that:

- Optimization of process parameters is a vital step that should be carried out carefully in order to achieve defect-free components with desired final characteristics.
- DED process parameters markedly affect the cooling rate, thermal gradient and, accordingly, thermal history and porosity content of the parts. It is well known that the quality of DED parts is chiefly determined by the process parameters, as well as the starting powder (particle size and chemical composition).
- Regarding the process parameters, the most important are laser power, scan speed, powder feed rate, building atmosphere, and deposition pattern. All these parameters influence the microstructure. The very high cooling rates of DED processes, with values around 10^3–10^4 °C/s, involve the formation of columnar and cellular structures based on the direction of thermal flux. It was reported that the columnar structures are dominant throughout the specimens, while the cellular structures are predominant in the last deposited layers.
- It is found that, the finer the PCAS, the higher the cooling rates. The high cooling rates generate very fine dendritic structures, as well as high dislocation densities, resulting in higher mechanical strength.
- The microstructure is composed of austenite γ and δ-ferrite, which is typically formed with the sub-grain structures enriched in Cr and Mo (δ-ferrite stabilize elements).
- Oxide formation is an undesired feature that affects the production of AISI 316L by the DED process. It is found that the presence of oxides can negatively affect the mechanical properties, even though an inert gas atmosphere is employed.
- The aforementioned microstructure features lead to materials with higher strength and lower ductility values with respect to conventionally processed AISI 316L stainless steel.
- Anisotropy in the tensile properties of DED components is widely detected; typically, the specimens produced along a direction parallel to the building platform present higher YS and US than specimens built perpendicular to the building platform. This can be attributed to different microstructure and thermal history, although there is a lack of extensive studies.

- Grain morphology, texture, elongated dendrites, and lack of fusion defects are found to be the main factors associated with the perpendicular building direction that have an effect on the tensile properties of DED AISI 316L components.
- Variations of the chemical composition associated with the recycling of the starting powder can influence microstructure and mechanical properties. In particular, the recycling of the powder can result in a higher oxide concentration (Mn and Si oxides) and, consequently, in a lower ductility of the final DED AISI 316L parts.

Author Contributions: To write this review article, A.S., A.A., and G.M. collected the papers, analyzed the literature, and wrote the article; S.B., M.L., and P.F. revised the article technically and scientifically. All authors have read and agreed to the published version of the manuscript.

Funding: This research received no external funding.

Conflicts of Interest: The authors declare no conflict of interest.

References

1. Saboori, A.; Gallo, D.; Biamino, S.; Fino, P.; Lombardi, M. An Overview of Additive Manufacturing of Titanium Components by Directed Energy Deposition: Microstructure and Mechanical Properties. *Appl. Sci.* **2017**, *7*, 883. [CrossRef]

2. Marchese, G.; Parizia, S.; Rashidi, M.; Saboori, A.; Manfredi, D.; Ugues, D.; Lombardi, M.; Hryha, E.; Biamino, S. The role of texturing and microstructure evolution on the tensile behavior of heat-treated Inconel 625 produced via laser powder bed fusion. *Mater. Sci. Eng. A* **2020**, *769*, 138500. [CrossRef]

3. Liou, F.; Slattery, K.; Kinsella, M.; Newkirk, J.W.; Landers, R.; Chou, H.-N. Applications of a hybrid manufacturing process for fabrication of metallic structures. *Rapid Prototyp. J.* **2007**, *13*, 236–244. [CrossRef]

4. Bosio, F.; Saboori, A.; Lacagnina, A.; Librera, E.; De Chirico, M.; Biamino, S.; Fino, P.; Lombardi, M. Directed energy deposition of 316L steel: Effect of type of powders and gas related parameters. In Proceedings of the Euro PM2018 Congress & Exhibition, Bilbao, Spain, 14–18 October 2018; pp. 1–6.

5. Galati, M.; Iuliano, L. A literature review of powder-based electron beam melting focusing on numerical simulations. *Addit. Manuf.* **2018**, *19*, 1–20. [CrossRef]

6. Aristizabal, M.; Jamshidi, P.; Saboori, A.; Cox, S.C.; Attallah, M.M. Laser powder bed fusion of a Zr-alloy: Tensile properties and biocompatibility. *Mater. Lett.* **2020**, *259*, 126897. [CrossRef]

7. Barros, R.; Silva, F.J.G.; Gouveia, R.; Saboori, A.; Marchese, G.; Biamino, S.; Salmi, A.; Atzeni, E. Laser Powder Bed Fusion of Inconel 718: Residual Stress Analysis Before and After Heat Treatment. *Metals* **2019**, *9*, 1290. [CrossRef]

8. Marchese, G.; Bassini, E.; Aversa, A.; Lombardi, M.; Ugues, D.; Fino, P.; Biamino, S. Microstructural Evolution of Post-Processed Hastelloy X Alloy Fabricated by Laser Powder Bed Fusion. *Materials* **2019**, *12*, 486. [CrossRef]

9. Frazier, W.E. Metal Additive Manufacturing: A Review. *J. Mater. Eng. Perform.* **2014**, *23*, 1917–1928. [CrossRef]

10. Del Guercio, G.; Galati, M.; Saboori, A.; Fino, P.; Iuliano, L. Microstructure and Mechanical Performance of Ti–6Al–4V Lattice Structures Manufactured via Electron Beam Melting (EBM): A Review. *Acta Met. Sin.* **2020**, *33*, 183–203. [CrossRef]

11. Mazzucato, F.; Valente, A.; Lai, M.; Biamino, S.; Lombardi, M.; Lombardi, M. Monitoring Approach to Evaluate the Performances of a New Deposition Nozzle Solution for DED Systems. *Technologies* **2017**, *5*, 29. [CrossRef]

12. Gibson, I.; Rosen, D.; Stucker, B. Directed Energy Deposition Processes. *Addit. Manuf. Technol.* **2015**, 245–268. [CrossRef]

13. Keicher, D.M.; Miller, W.D. LENSTM moves beyond RP to direct fabrication. *Met. Powder Rep.* **1998**, *53*, 26–28.

14. Wołosz, P.; Baran, A.; Polański, M. The influence of laser engineered net shaping (LENS^{TM}) technological parameters on the laser deposition efficiency and properties of H13 (AISI) steel. *J. Alloys Compd.* **2020**, *823*, 153840. [CrossRef]

15. Petrat, T.; Graf, B.; Gumenyuk, A.; Rethmeier, M. Laser Metal Deposition as Repair Technology for a Gas Turbine Burner Made of Inconel 718. *Phys. Procedia* **2016**, *83*, 761–768. [CrossRef]

16. Zhang, Y.; Yang, L.; Lu, W.; Wei, D.; Meng, T.; Gao, S. Microstructure and elevated temperature mechanical properties of IN718 alloy fabricated by laser metal deposition. *Mater. Sci. Eng. A* **2020**, *771*, 138580. [CrossRef]

17. Weerasinghe, V.M.; Steen, W.M. Laser Cladding By Powder Injection. In Proceedings of the 1st International Conference on Lasers in Manufacturing, Brighton, UK, 1–3 November 1983; pp. 125–132.

18. Weerasinghe, V.M.; Steen, W.M. Laser Cladding With Blown Powder. *Met. Constr.* **1987**, *19*, 581–585.

19. Mazumder, J.; Choi, J.; Nagarathnam, K.; Koch, J.; Hetzner, D. The direct metal deposition of H13 tool steel for 3-D components. *JOM* **1997**, *49*, 55–60. [CrossRef]

20. Chen, B.; Su, Y.; Xie, Z.; Tan, C.; Feng, J. Development and characterization of 316L/Inconel625 functionally graded material fabricated by laser direct metal deposition. *Opt. Laser Technol.* **2020**, *123*, 105916. [CrossRef]

21. Zhang, J.; Liou, F. Adaptive Slicing for a Multi-Axis Laser Aided Manufacturing Process. *J. Mech. Des.* **2004**, *126*, 254–261. [CrossRef]

22. Milewski, J.; Lewis, G.; Thoma, D.; Keel, G.; Nemec, R.; Reinert, R. Directed light fabrication of a solid metal hemisphere using 5-axis powder deposition. *J. Mater. Process. Technol.* **1998**, *75*, 165–172. [CrossRef]

23. Wu, X.; Liang, J.; Mei, J.; Mitchell, C.; Goodwin, P.; Voice, W. Microstrucfures of laser-deposited Ti–6Al–4V. *Mater. Des.* **2004**, *25*, 137–144. [CrossRef]

24. Shao, S.; Yadollahi, A.; Bian, L.; Thompson, S.M. An overview of Direct Laser Deposition for additive manufacturing; Part II: Mechanical behavior, process parameter optimization and control. *Addit. Manuf.* **2015**, *8*, 12–35. [CrossRef]

25. Saboori, A.; Aversa, A.; Marchese, G.; Biamino, S.; Lombardi, M.; Fino, P. Application of Directed Energy Deposition-Based Additive Manufacturing in Repair. *Appl. Sci.* **2019**, *9*, 3316. [CrossRef]

26. Mazumder, J.; Dutta, D.; Kikuchi, N.; Ghosh, A. Closed loop direct metal deposition: Art to part. *Opt. Lasers Eng.* **2000**, *34*, 397–414. [CrossRef]

27. Bontha, S.; Klingbeil, N.; Kobryn, P.A.; Fraser, H.L. Thermal process maps for predicting solidification microstructure in laser fabrication of thin-wall structures. *J. Mater. Process. Technol.* **2006**, *178*, 135–142. [CrossRef]

28. Bontha, S.; Klingbeil, N.; Kobryn, P.A.; Fraser, H.L. Effects of process variables and size-scale on solidification microstructure in beam-based fabrication of bulky 3D structures. *Mater. Sci. Eng. A* **2009**, *513*, 311–318. [CrossRef]

29. Zadi-Maad, A.; Rohib, R.; Irawan, A. Additive manufacturing for steels: A review. *IOP Conf. Series: Mater. Sci. Eng* **2018**, *285*, 12028. [CrossRef]

30. Saboori, A.; Tusacciu, S.; Busatto, M.; Lai, M.; Biamino, S.; Fino, P.; Lombardi, M. Production of Single Tracks of Ti-6Al-4V by Directed Energy Deposition to Determine the Layer Thickness for Multilayer Deposition. *J. Vis. Exp.* **2018**, e56966. [CrossRef]

31. Wang, Z.; Palmer, T.A.; Beese, A.M. Effect of processing parameters on microstructure and tensile properties of austenitic stainless steel 304L made by directed energy deposition additive manufacturing. *Acta Mater.* **2016**, *110*, 226–235. [CrossRef]

32. Majumdar, J.D.; Pinkerton, A.; Liu, Z.; Manna, I.; Li, L. Mechanical and electrochemical properties of multiple-layer diode laser cladding of 316L stainless steel. *Appl. Surf. Sci.* **2005**, *247*, 373–377. [CrossRef]

33. Sun, G.F.; Shen, X.; Wang, Z.; Zhan, M.; Yao, S.; Zhou, R.; Ni, Z. Laser metal deposition as repair technology for 316L stainless steel: Influence of feeding powder compositions on microstructure and mechanical properties. *Opt. Laser Technol.* **2019**, *109*, 71–83. [CrossRef]

34. Bertoli, U.S.; Guss, G.; Wu, S.; Matthews, M.; Schoenung, J.M. In-situ characterization of laser-powder interaction and cooling rates through high-speed imaging of powder bed fusion additive manufacturing. *Mater. Des.* **2017**, *135*, 385–396. [CrossRef]

35. Ziętala, M.; Durejko, T.; Polański, M.; Kunce, I.; Płociński, T.; Zieliński, W.; Łazińska, M.; Stępniowski, W.; Czujko, T.; Kurzydłowski, K.J.; et al. The microstructure, mechanical properties and corrosion resistance of 316L stainless steel fabricated using laser engineered net shaping. *Mater. Sci. Eng. A* **2016**, *677*, 1–10. [CrossRef]

36. Yadollahi, A.; Seely, D.; Patton, B.; Shao, S. Microstructural Features and Mechanical Properties of 316L Stainless Steel fabricated by Laser Additive Manufacture. In Proceedings of the 56th AIAA/ASCE/AHS/ASC Structures, Structural Dynamics, and Materials Conference, Kissimmee, FL, USA, 9 January 2015. [CrossRef]

37. Saboori, A.; Aversa, A.; Bosio, F.; Bassini, E.; Librera, E.; De Chirico, M.; Biamino, S.; Ugues, D.; Fino, P.; Lombardi, M. An investigation on the effect of powder recycling on the microstructure and mechanical properties of AISI 316L produced by Directed Energy Deposition. *Mater. Sci. Eng. A* **2019**, *766*, 138360. [CrossRef]

38. Saboori, A.; Piscopo, G.; Lai, M.; Salmi, A.; Biamino, S. An investigation on the effect of deposition pattern on the microstructure, mechanical properties and residual stress of 316L produced by Directed Energy Deposition. *Mater. Sci. Eng. A* **2020**, *780*, 139179. [CrossRef]

39. Zheng, B.; Haley, J.; Yang, N.; Yee, J.; Terrassa, K.; Zhou, Y.; Lavernia, E.; Schoenung, J. On the evolution of microstructure and defect control in 316L SS components fabricated via directed energy deposition. *Mater. Sci. Eng. A* **2019**, *764*, 138243. [CrossRef]

40. Terrassa, K.L.; Smith, T.R.; Jiang, S.; Sugar, J.D.; Schoenung, J.M. Improving build quality in Directed Energy Deposition by cross-hatching. *Mater. Sci. Eng. A* **2019**, *765*, 138269. [CrossRef]

41. Tan, Z.E.; Pang, J.H.L.; Kaminski, J.; Pepin, H.; Zhi'En, E.T. Characterisation of porosity, density, and microstructure of directed energy deposited stainless steel AISI 316L. *Addit. Manuf.* **2019**, *25*, 286–296. [CrossRef]

42. Griffith, M.; Schlienger, M.; Harwell, L.; Oliver, M.; Baldwin, M.; Ensz, M.; Essien, M.; Brooks, J.; Robino, C.; Smugeresky, J.; et al. Understanding thermal behavior in the LENS process. *Mater. Des.* **1999**, *20*, 107–113. [CrossRef]

43. Kobryn, P.A.; Semiatin, S.L. Mechanical properties of laser-deposited Ti-6Al-4V. In Proceedings of the Solid Freeform fabrication proceedings, Austin, TX, USA, 6 August 2001; pp. 179–186.

44. Saboori, A.; Biamino, S.; Lombardi, M.; Tusacciu, S.; Busatto, M.; Lai, M.; Fino, P. How the nozzle position affects the geometry of the melt pool in directed energy deposition process. *Powder Met.* **2019**, *62*, 213–217. [CrossRef]

45. Selcuk, C. Laser metal deposition for powder metallurgy parts. *Powder Met.* **2011**, *54*, 94–99.

46. Vilar, R. Laser cladding. *Laser Appl.* **2001**, *11*, 64–79. [CrossRef]

47. Ensz, M.; Griffith, M.; Hofmeister, W.; Philliber, J.A.; Smugeresky, J.; Wert, M. *Investigation of Solidification in the Laser Engineered Net Shaping (LENS) Process*; Sandia National Laboratories: Livermore, CA, USA, 1999.

48. Debroy, T.; Wei, H.; Zuback, J.; Mukherjee, T.; Elmer, J.; Milewski, J.; Beese, A.M.; Wilson-Heid, A.; De, A.; Zhang, W. Additive manufacturing of metallic components—Process, structure and properties. *Prog. Mater. Sci.* **2018**, *92*, 112–224. [CrossRef]

49. Ma, M.; Wang, Z.; Zeng, X. A comparison on metallurgical behaviors of 316L stainless steel by selective laser melting and laser cladding deposition. *Mater. Sci. Eng. A* **2017**, *685*, 265–273. [CrossRef]

50. El Kadiri, H.; Wang, L.; Horstemeyer, M.F.; Yassar, R.S.; Berry, J.T.; Felicelli, S.; Wang, P.T. Phase transformations in low-alloy steel laser deposits. *Mater. Sci. Eng. A* **2008**, *494*, 10–20. [CrossRef]

51. Zheng, B.; Zhou, Y.; Smugeresky, J.; Schoenung, J.; Lavernia, E. Thermal Behavior and Microstructural Evolution during Laser Deposition with Laser-Engineered Net Shaping: Part I. Numerical Calculations. *Met. Mater. Trans. A* **2008**, *39*, 2228–2236. [CrossRef]

52. Bi, G.; Gasser, A.; Wissenbach, K.; Drenker, A.; Poprawe, R. Characterization of the process control for the direct laser metallic powder deposition. *Surf. Coat. Technol.* **2006**, *201*, 2676–2683. [CrossRef]

53. Kurz, W. Solidification Microstructure-Processing Maps: Theory and Application. *Adv. Eng. Mater. Banner.* **2001**, *3*, 443–452. [CrossRef]

54. Kelly, S.M.; Kampe, S. Microstructural evolution in laser-deposited multilayer Ti-6Al-4V builds: Part I. Microstructural characterization. *Met. Mater. Trans. A* **2004**, *35*, 1861–1867. [CrossRef]

55. Colaço, R.; Vilar, R. Phase selection during laser surface melting of martensitic stainless tool steels. *Scr. Mater.* **1997**, *36*, 199–205. [CrossRef]

56. Hofmeister, W.; Griffith, M. Solidification in direct metal deposition by LENS processing. *JOM* **2001**, *53*, 30–34. [CrossRef]

57. Griffith, M.L. Understanding the microstructure and properties of components fabricated by laser engineered net shaping (LENS^{TM}). In Proceedings of the Materials Research Society Symposium, Boston, MA, USA, 26 November–1 December 2000; p. 625.

58. Grujicic, M.; Cao, G.; Figliola, R. Computer simulations of the evolution of solidification microstructure in the LENS™ rapid fabrication process. *Appl. Surf. Sci.* **2001**, *183*, 43–57. [CrossRef]

59. Ye, R.; Smugeresky, J.E.; Zheng, B.; Zhou, Y.; Lavernia, E.J. Numerical modeling of the thermal behavior during the LENS® process. *Mater. Sci. Eng. A* **2006**, *428*, 47–53. [CrossRef]

60. Hofmeister, W.; Wert, M.; Smugeresky, J.; Philliber, J.A.; Griffith, M.; Ensz, M. Investigating Solidification with the Laser-Engineered Net Shaping (LENSTM) Process. *JOM* **1999**, *51*, 1–6.

61. Ghosh, S.; Choi, J. Modeling and Experimental Verification of Transient/Residual Stresses and Microstructure Formation in Multi-Layer Laser Aided DMD Process. *J. Heat Transf.* **2005**, *128*, 662–679. [CrossRef]

62. Yin, H.; Felicelli, S.D. Dendrite growth simulation during solidification in the LENS process. *Acta Mater.* **2010**, *58*, 1455–1465. [CrossRef]

63. Flemings, M.C. Solidification processing. *Met. Mater. Trans. A* **1974**, *5*, 2121–2134. [CrossRef]

64. Aversa, A.; Saboori, A.; Librera, E.; de Chirico, M.; Biamino, S.; Lombardi, M.; Fino, P. The Role of Directed Energy Deposition Atmosphere Mode on the Microstructure and Mechanical Properties of 316L Samples. *Addit. Manuf.* **2020**, 101274. [CrossRef]

65. Song, J.; Deng, Q.; Chen, C.; Hu, D.; Li, Y. Rebuilding of metal components with laser cladding forming. *Appl. Surf. Sci.* **2006**, *252*, 7934–7940. [CrossRef]

66. Syed, W.U.H.; Pinkerton, A.; Li, L. A comparative study of wire feeding and powder feeding in direct diode laser deposition for rapid prototyping. *Appl. Surf. Sci.* **2005**, *247*, 268–276. [CrossRef]

67. Smugeresky, J.; Keicher, D.; Romero, J.; Griffith, M.; Harwell, L. aser engineered net shaping(LENS) process: Optimization of surface finish and microstructural properties. *Adv. Powder Metall. Part. Mater.* **1997**, *3*, 21.

68. Kocabekir, B.; Kaçar, R.; Gündüz, S.; Hayat, F. An effect of heat input, weld atmosphere and weld cooling conditions on the resistance spot weldability of 316L austenitic stainless steel. *J. Mater. Process. Technol.* **2008**, *195*, 327–335. [CrossRef]

69. Jacob, G. Prediction of Solidification Phases in Cr-Ni Stainless Steel Alloys Manufactured by Laser Based Powder Bed Fusion Process, NIST. *Adv. Manuf. Ser.* **2018**, *100–114*, 1–38.

70. Guo, P.; Zou, B.; Huang, C.; Gao, H. Study on microstructure, mechanical properties and machinability of efficiently additive manufactured AISI 316L stainless steel by high-power direct laser deposition. *J. Mater. Process. Technol.* **2017**, *240*, 12–22. [CrossRef]

71. De Lima, M.S.F.; Sankare, S. Microstructure and mechanical behavior of laser additive manufactured AISI 316 stainless steel stringers. *Mater. Des.* **2014**, *55*, 526–532. [CrossRef]

72. Saboori, A.; Toushekhah, M.; Aversa, A.; Lai, M.; Lombardi, M.; Biamino, S.; Fino, P. Critical Features in the Microstructural Analysis of AISI 316L Produced By Metal Additive Manufacturing. *Met. Microstruct. Anal.* **2020**, *9*, 92–96. [CrossRef]

73. Yan, F.; Xiong, W.; Faierson, E.; Olson, G. Characterization of nano-scale oxides in austenitic stainless steel processed by powder bed fusion. *Scr. Mater.* **2018**, *155*, 104–108. [CrossRef]

74. Babu, S.S.; David, S.A.; Vitek, J.M.; Mundra, K.; Debroy, T. Development of macro- and microstructures of carbon–manganese low alloy steel welds: Inclusion formation. *Mater. Sci. Technol.* **1995**, *11*, 186–199. [CrossRef]

75. Lou, X.; Andresen, P.L.; Rebak, R.B. Oxide inclusions in laser additive manufactured stainless steel and their effects on impact toughness and stress corrosion cracking behavior. *J. Nucl. Mater.* **2018**, *499*, 182–190. [CrossRef]

76. Ganesh, P.; Giri, R.; Kaul, R.; Sankar, P.R.; Tiwari, P.; Atulkar, A.; Porwal, R.; Dayal, R.; Kukreja, L. Studies on pitting corrosion and sensitization in laser rapid manufactured specimens of type 316L stainless steel. *Mater. Des.* **2012**, *39*, 509–521. [CrossRef]

77. Saboori, A.; Dadkhah, M.; Fino, P.; Pavese, M. An Overview of Metal Matrix Nanocomposites Reinforced with Graphene Nanoplatelets; Mechanical, Electrical and Thermophysical Properties. *Metals* **2018**, *8*, 423. [CrossRef]

78. Saboori, A.; Pavese, M.; Badini, C.; Fino, P. Effect of Sample Preparation on the Microstructural Evaluation of Al–GNPs Nanocomposites. *Met. Microstruct. Anal.* **2017**, *6*, 619–622. [CrossRef]

79. Eo, D.-R.; Park, S.-H.; Cho, J.-W. Inclusion evolution in additive manufactured 316L stainless steel by laser metal deposition process. *Mater. Des.* **2018**, *155*, 212–219. [CrossRef]

80. Yadollahi, A.; Shao, S.; Thompson, S.M.; Seely, D.W. Effects of process time interval and heat treatment on the mechanical and microstructural properties of direct laser deposited 316L stainless steel. *Mater. Sci. Eng. A* **2015**, *644*, 171–183. [CrossRef]

81. Gale, J.; Achuhan, A. Application of ultrasonic peening during DMLS production of 316L stainless steel and its effect on material behavior. *Rapid Prototyp. J.* **2017**, *23*, 1185–1194. [CrossRef]

82. Saeidi, K.; Gao, X.; Lofaj, F.; Kvetková, L.; Shen, Z. Transformation of austenite to duplex austenite-ferrite assembly in annealed stainless steel 316L consolidated by laser melting. *J. Alloys Compd.* **2015**, *633*, 463–469. [CrossRef]

83. Zhang, K.; Wang, S.; Liu, W.; Shang, X. Characterization of stainless steel parts by Laser Metal Deposition Shaping. *Mater. Des.* **2014**, *55*, 104–119. [CrossRef]

84. Yang, N.; Yee, J.; Zheng, B.; Gaiser, K.; Reynolds, T.; Clemon, L.; Lu, W.Y.; Schoenung, J.M.; Lavernia, E.J. Process-Structure-Property Relationships for 316L Stainless Steel Fabricated by Additive Manufacturing and Its Implication for Component Engineering. *J. Spray Technol.* **2016**, *26*, 610–626. [CrossRef]

85. Saeidi, K.; Gao, X.; Zhong, Y.; Shen, Z. Hardened austenite steel with columnar sub-grain structure formed by laser melting. *Mater. Sci. Eng. A* **2015**, *625*, 221–229. [CrossRef]

86. Kim, N.-K.; Woo, W.; Kim, E.-Y.; Choi, S.-H. Microstructure and mechanical characteristics of multi-layered materials composed of 316L stainless steel and ferritic steel produced by direct energy deposition. *J. Alloys Compd.* **2019**, *774*, 896–907. [CrossRef]

87. Park, J.S.; Park, J.H.; Lee, M.-G.; Sung, J.H.; Cha, K.J.; Kim, D.H. Effect of Energy Input on the Characteristic of AISI H13 and D2 Tool Steels Deposited by a Directed Energy Deposition Process. *Met. Mater. Trans. A* **2016**, *47*, 2529–2535. [CrossRef]

88. Gülenç, B.; Develi, K.; Kahraman, N.; Durgutlu, A. Experimental study of the effect of hydrogen in argon as a shielding gas in MIG welding of austenitic stainless steel. *Int. J. Hydrogen Energy* **2005**, *30*, 1475–1481. [CrossRef]

89. Boes, J.; Röttger, A.; Becker, L.; Theisen, W. Processing of gas-nitrided AISI 316L steel powder by laser powder bed fusion—Microstructure and properties. *Addit. Manuf.* **2019**, *30*, 100836. [CrossRef]

90. Morrow, B.; Lienert, T.J.; Knapp, C.M.; Sutton, J.O.; Brand, M.J.; Pacheco, R.M.; Livescu, V.; Carpenter, J.S.; Gray, G.T. Impact of Defects in Powder Feedstock Materials on Microstructure of 304L and 316L Stainless Steel Produced by Additive Manufacturing. *Met. Mater. Trans. A* **2018**, *49*, 3637–3650. [CrossRef]

91. Ganesh, P.; Kaul, R.; Sasikala, G.; Kumar, H.; Venugopal, S.; Tiwari, P.; Rai, S.; Prasad, R.C.; Kukreja, L.M. Fatigue Crack Propagation and Fracture Toughness of Laser Rapid Manufactured Structures of AISI 316L Stainless Steel. *Met. Microstruct. Anal.* **2014**, *3*, 36–45. [CrossRef]

92. Gordon, J.; Hochhalter, J.; Haden, C.; Harlow, D.G. Enhancement in fatigue performance of metastable austenitic stainless steel through directed energy deposition additive manufacturing. *Mater. Des.* **2019**, *168*, 107630. [CrossRef]

applied sciences

Article

Dense, Strong, and Precise Silicon Nitride-Based Ceramic Parts by Lithography-Based Ceramic Manufacturing

Altan Alpay Altun [1], Thomas Prochaska [1], Thomas Konegger [2] and Martin Schwentenwein [1,*]

[1] Lithoz GmbH, Mollardgasse 85A/2/64-69, 1060 Vienna, Austria; aaltun@lithoz.com (A.A.A.); tprochaska@lithoz.com (T.P.)

[2] TU Wien, Institute of Chemical Technologies and Analytics, Getreidemarkt 9/164-CT, 1060 Vienna, Austria; thomas.konegger@tuwien.ac.at

* Correspondence: mschwentenwein@lithoz.com; Tel.: +43-1-934-661-2204

Received: 27 December 2019; Accepted: 25 January 2020; Published: 3 February 2020

Abstract: Due to the high level of light absorption and light scattering of dark colored powders connected with the high refractive indices of ceramic particles, the majority of ceramics studied via stereolithography (SLA) have been light in color, including ceramics such as alumina, zirconia and tricalcium phosphate. This article focuses on a lithography-based ceramic manufacturing (LCM) method for β-SiAlON ceramics that are derived from silicon nitride and have excellent material properties for high temperature applications. This study demonstrates the general feasibility of manufacturing of silicon nitride-based ceramic parts by LCM for the first time and combines the advantages of SLA, such as the achievable complexity and low surface roughness ($R_a = 0.50$ μm), with the typical properties of conventionally manufactured silicon nitride-based ceramics, such as high relative density (99.8%), biaxial strength ($σf = 764$ MPa), and hardness ($HV_{10} = 1500$).

Keywords: additive manufacturing; silicon nitride; high performance ceramics; photopolymerisation; lithography-based ceramic manufacturing

---

1. Introduction

The history of silicon nitride (Si_3N_4) ceramics began about 150 years ago, with Deville and Wöhler developing silicon nitride synthetically in 1857 [1], even though naturally occurring nierite minerals, such as $α-Si_3N_4$ and $β-Si_3N_4$, have been found during later, detailed analyses of particles of meteoritic rock [2]. Silicon nitride currently plays a dominant role in the field of nonoxide ceramics, and exhibits a combination of excellent material properties, such as high toughness and strength at high temperatures, excellent thermal shock resistance and good chemical resistance, which is unmatched by other ceramics [3]. Due to these superior material properties, silicon nitride ceramics became popular in the 1950s, e.g., for use in thermocouple tubing [4]. Owing to these properties, in particular thermal conductivity and thermal and corrosion resistance, silicon nitride-based ceramics can also be used as base material for thermal conductors, gas turbines, and ball bearings [5]. Apart from technical applications silicon nitride-based ceramics have also been used as material for medical devices (e.g., spinal cages). It was shown that the material shows good osseointegration and stimulated cell differentiation as well as osteoblastic activity which resulted in accelerated bone ingrowths compared to poly (ether ether ketone) (PEEK) [6]. Furthermore, silicon nitride was also evaluated regarding its anti-infective behavior when used for implants. Webster et al. compared the bacterial growth on calvarial implants in rats of silicon nitride to implants made of either titanium or PEEK [7]. They observed bacteria in 88% of the PEEK implants and 12% of the titanium implants whereas no bacteria were present adjacent to the silicon nitride implants. Similar results have also been observed by

Ishikawa et al. in tibial implants [8]. These properties, together with its excellent biocompatibility and mechanical strength, make silicon nitride to a very attractive material for biomedical engineering [9]. The combination of load-bearing ability and induction of cell differentiation makes it superior to classical bone replacement materials like calcium phosphates. Additionally, the bactericidal behavior could drastically reduce infection associated implant failures or necessary revision surgeries, which have quadrupled in the last 20 years [10].

The solid phase sintering of pure silicon nitride ceramics is generally not feasible due to decomposition processes and low diffusion coefficients. Based on that reasoning, sintering additives are needed to allow for consolidation by sintering. The first sintering aid investigated for the densification of Si_3N_4 ceramics was magnesium oxide (MgO). Subsequently, other materials such as alumina (Al_2O_3), yttria (Y_2O_3), zirconia (ZrO_2) and ceria (CeO_2) were used to further promote densification. Overall, the yttria–alumina (Y_2O_3-Al_2O_3) system is the most widely applied sintering aid system [11]. An effective method for increasing the creep and oxidation resistance of silicon nitride is the addition of large amounts of sintering additives to the silicon nitride structure in the grain boundary phase, the formed mixed crystals being known as SiAlONs [12,13]. SiAlON ceramics are the solid solution of silicon nitride and the sintering additives alumina (Al_2O_3), yttria (Y_2O_3) and aluminum nitride (AlN). In the early 1970s, the formation of this solid solution that has the same structural form as silicon nitride has been reported in Japan and in the UK parallel [14]. Since the SiAlON powders contain the sintering aids, the sintering additives were not blended additionally in this work.

Conventional methods for manufacturing intricate ceramic bodies demand difficult and costly production steps. As a result, the manufacturing industry becomes less interested in creating highly complex geometries due to the high price of the molds, the extended production times, and the challenging consolidation behavior. However, with the introduction of additive manufacturing (AM), there was new potential to create elaborate shapes which had been impossible to achieve using conventional techniques. AM refers to a class of technologies in which a three dimensional (3D) object is manufactured directly from a computer aided design (CAD) model. It is defined by ASTM F2792—12a (Standard Terminology for Additive Technologies) as "the process of joining materials to make objects from 3D model data, usually layer upon layer, as opposed to subtractive manufacturing, such as traditional machining" [15].

Ceramic AM technologies can be divided into two groups: direct and indirect methods [16]. Typical steps involved in indirect methods include casting of ceramic suspension into a temporary mold, which is fabricated via conventional 3D printing (3DP), followed by the subsequent solidifying and sintering process. The main advantages of indirect methods of AM are the absence of delamination processes and the isotropy of characteristics; however, problems arise when discarding the mold, creating a possible dilemma when using indirect methods. Direct methods, on the other hand, allow for more freedom in terms of complexity, and have less processing steps due to the fabrication of ceramic materials, layer-by-layer, directly onto the building platform. Depending on the raw materials, direct methods can also be divided into powder-, suspension-, precursor-, melting-, and reaction-based processes. Powder-based processes, such as binder-jet 3DP and selective laser melting (powder bed fusion), use ceramic particles in a powder bed to build 3D objects. The advantage of powder bed fusion is the possibility to manufacture ceramics parts without requiring any subsequent sintering process, however, the fabricated parts can be porous and the thermal distortion can cause warping of fabricated objects [17–19]. In the suspension-based processes, ceramic powders are formed with the aid of a matrix of monomers, polymers, photopolymers, or solvents. Stereolithography (SLA), laminated object manufacturing (LOM), robocasting and fused filament fabrication (FFF) are the typical examples of processes depending on the type of binder [20–23]. The process is completed by removing the organic components through evaporation or decomposition and subsequent densification by sintering.

Among these techniques, SLA is still the most prominent AM method due to the high resolution of the surface, facilitating the production of an increasing number of different types of materials. Precursor-based processes rely on the conversion of inorganic polymers into ceramics through pyrolysis.

The limited thickness of these materials due to the extremely high volume loss during the pyrolysis step is the main challenges here [24]. Melting-based processes can only be used with low melting point ceramics such as glass. Melted glass is deposited through a nozzle and then cooled to room temperature in order to obtain transparent materials with superior material properties [25]. Reaction-based processes (e.g. plaster-based 3D inkjet printing), in contrast, generally require thermal post-processing treatment to obtain the desired final materials. Recently a modified technique has been reported involving a chemical reaction of ceramic particles with an ink to fabricate ceramic bodies without requiring any further thermal post-processing treatments such as sintering [16].

Lithography-Based Ceramic Manufacturing Process

Lithography-based techniques are based upon the concept of photopolymerisation. The lithography-based ceramic manufacturing process has been developed for highly filled and highly viscous ceramic suspension. Figure 1 shows the schematic system of the LCM process. A photosensitive formulation is cured in the required areas through selective light exposure to light. In the case of lithography-based ceramic manufacturing (LCM), the suspension is cured via a mask-exposure process using the concept of digital light processing (DLP). The optics used currently generate a pixel size of 40 × 40 μm with a resolution of 1920 × 1080 pixels.

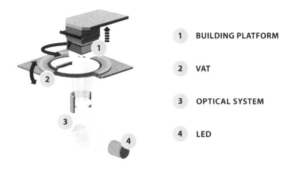

Figure 1. Schematic drawing of the Lithography-based Ceramic Manufacturing (LCM) process: (1) building platform; (2) vat; (3) optical system; (4) light engine.

Once the three-dimensional structure is shaped, the part is called a green body, in which the ceramic particles are now surrounded by a polymer network; however, the photopolymer only acts as a binder in the ceramic green body [26]. For this reason, this composite material requires additional thermal treatment through the processes of debinding and sintering. During the debinding step, the polymer network is burned off and the subsequent sintering step causes densification by fusing the ceramic particles together.

The LCM process is carried out using a printer composed of a building platform, a vat, an optical system and a light engine (Figure 1). First, the vat is filled with the photocurable ceramic suspension and the building platform then begins to move down into the suspension until the gap between the building platform and the vat is a chosen distance, typically between 10 and 100 μm. This gap, corresponding to the resulting thickness of the printed layer of the green body, is chosen according to the optical properties and photoreactivity of the ceramic suspension as well as the needed resolution of the printed part. Then, the photocurable suspension is cured selectively through a mask-exposure process from the bottom of the transparent vat. The light engine is based on light-emitting diodes (LED) with a wavelength of 460 nm. By repeating this process layer for layer, the green body is manufactured.

After printing of the green body, the LCM process also involves a thermal treatment which comprises debinding (removal of binder) and sintering (densification of ceramics), comparable to conventional techniques (see Figure 2) [26]. The removal of organic components is the most critical step in the process. For this reason, a reduction of binder concentration or a high solids loading of ceramic particles is desired. During the debinding, the photocured binder is burned off, typically under air atmosphere. Due to the evaporation or decomposition behavior of organic components at different temperatures, a tailored temperature profile has to be used to obtain crack-free ceramics after thermal post processing [27]. Here, thermogravimetric analysis and differential scanning calorimeter are suitable tools to select adequate parameters. In contrast to oxide ceramics, nonoxide ceramics, such as silicon nitride and silicon carbide, need an inert atmosphere during sintering step in order to avoid oxidation. The interaction between ceramics particles at high temperatures induces densification via sintering and, consequently, fully dense ceramics are formed [28,29].

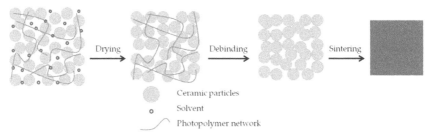

Figure 2. Steps of the thermal treatment after green body fabrication: drying, debinding and sintering.

LCM has been demonstrated to be highly capable when it comes to the precision and mechanical properties of the fabricated parts and has already been successfully applied to fabricate complex-shaped ceramics from alumina, yttria-stabilized zirconia, tricalcium phosphate, mullite, silicon oxycarbide and magnesia [30–35].

In this work, we report on the development of an LCM process for the production of highly complex-shaped silicon nitride-based ceramics exhibiting mechanical properties at the same level of conventionally manufactured materials. As processing of silicon nitride powders by stereolithographic methods is highly challenging, both during the AM stage and during the consolidation stage, this has to our knowledge not been achieved previously, our study thus being the first of its kind. Chung et al. have published that the silicon nitride-based green bodies could be printed by digital light processing; however, the sintered silicon nitride-based ceramics have a relative density of approximately 90% which is significantly lower than the results obtained within this work [36]. In addition to this, the published article has a lack of some analyses such as surface roughness of printed samples, stability of the ceramic suspension, thermogravimetric analysis of the ceramic suspension. According to the photos from Chung' work, the resolution and complexity of the printed samples are also incomparably lower than the results obtained within this work.

2. Materials and Methods

2.1. Suspension Characterization

In the LCM process, the layer thickness is one of the crucial parameters for the shaping process. For this reason, light penetration tests were conducted to quantify how far light can penetrate into a given ceramic suspension. The ceramic suspension was exposed to the same level of maximum light intensity (47.1 mW cm^{-2}) for five different time periods, resulting in distinct energy doses through a cylinder with a diameter of 10 mm. The number of tested suspension samples was 3 for each of the energy doses. The thickness of the polymerized layer was then measured using a micrometer screw. Once the thickness of the polymerized layer was determined, the lateral over-polymerization was also measured for all polymerized layer with an aid of a light microscope. The difference between target value and actual value was noted as over-polymerization. Griffith et al. have shown that the scattering efficiency term (Q) is proportional to the difference of refractive indices (n) between the ceramic particles and photocurable matrix.

$$Q = \beta \, \Delta n^2 \tag{1}$$

$$\Delta n^2 = (n_{ceramic} - n_{solution})^2 \tag{2}$$

The penetration depth is inversely proportional to the refractive index, and the term β is relevant to the particle sizes and wavelengths [26]. The refractive index of silicon nitride powder is 2.0167 [37], whereas the refractive index of organic binder is 1.4630 ($\pm$ 0.0005).

The rheological measurements of photocurable suspension were performed on a rheometer (MCR 92, Anton Paar, Graz, Austria) at a temperature of 20 °C with a plate–plate arrangement. The plate–plate measuring system has a diameter of 25 mm and the gap between plates was 0.5 mm. The viscosity was measured three times for each linearly at shear rates between 5 and 50 s^{-1}.

Finally, an evaluation of the typical materials efficiency was performed for the LCM process in combination with the LithaNit 720 suspension that was employed in this work. For this reason, the weight of the suspension was recorded both before and after printing. Furthermore, the exact mass of printed samples in the green state was measured once the parts had been cleaned. The factor of suspension was calculated by dividing the mass of the suspension for printing by the mass of the printed parts after cleaning is a measure for the materials efficiency of the used LCM process. This process was repeated three times in order to find a mean value and standard deviation.

2.2. Sample Preparation

In this study, the silicon nitride-based ceramic parts were manufactured using a commercially available photocurable ceramic suspension LithaNit 720 (Lithoz GmbH, Vienna, Austria), which has a solid loading of 40 vol. %. LithaNit 720 consists of a SiAlON powder blend and a photocurable binder system.

After printing and cleaning of the samples, debinding was conducted in an air furnace (HTCT 08/16, Nabertherm, Lilienthal, Germany), using a temperature profile as shown in Figure 3. Subsequently, the debinded samples were transferred to the sintering furnace. Sintering was performed at 1750 °C with a dwelling time of 5 hours under nitrogen atmosphere using a graphite-heated furnace (KCE HP W 150/200-2200-100 LA). During sintering, the samples were embedded in a powder bed consisting of a mixture of silicon nitride and boron nitride as sintering aids to ensure the separation of the sintered ceramic parts and crucibles.

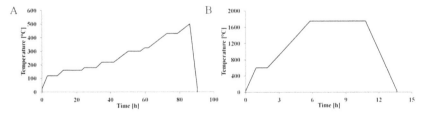

Figure 3. Complex temperature profile of silicon nitride-based green bodies: (**a**) debinding (**b**) sintering.

2.3. Sample Characterization

Thermogravimetric analysis (TGA 2050; TA Instruments, New Castle, USA) was performed to quantify the mass loss during a thermal treatment as a function of temperature, using a cylindrical sample with both a diameter and height of 6 mm. The specimen was heated from room temperature to 500 °C at a rate of 2 K/min in nitrogen atmosphere and air.

The surface roughness of the printed silicon nitride-based ceramics was analyzed by tactile profilometry using a Surftest SJ-400 Profilometer (Mitutoyo, Kanagawa, Japan) according to ISO

(International Organization for Standardization) 1997. Figure 4 depicts the orientations of the samples on the building platform of the CeraFab machine (a) and the sintered ceramic samples (b). The sintered ceramic parts have a main dimension of 14.66 × 2.00 × 5.54 mm (L × W × H). The aim of three different orientations such as X, Y and 45° is to see the effect of the projector pixel pattern and to interpret the results statistically. As such, the assessed samples were tested post-firing without subsequent surface treatment such as grinding or polishing.

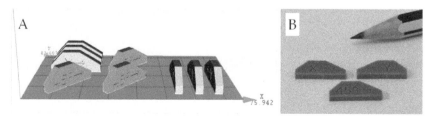

Figure 4. Samples for surface roughness measurement: (**a**) 3D model (**b**) sintered samples.

Density measurements were conducted for three cylinders with a diameter of 6 mm and a height of 10 mm, following the Archimedean principle using a density determination kit (SI-234A, Denver Instrument, Bohemia, NY, USA), allowing for determination of relative density (assuming a theoretical density of 3.24 g cm^{-3}).

The biaxial flexural test was used to characterize the mechanical strength of the brittle silicon nitride-based ceramic samples according to ASTM (American Society for Testing and Materials) F394-78. The samples had a diameter of 22 mm and a thickness of 2.5 mm. The number of tested samples per batch was 7. Before testing the samples, the disc surfaces were polished. The tests were performed on a universal testing machine (Instron Universal, MA, USA). As shown in Figure 5a, the specimen was placed on supporting balls and the central force was applied by a fourth loading ball.

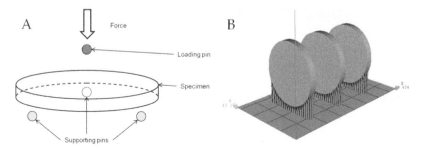

Figure 5. (**a**) Principle of biaxial flexural strength measurement; (**b**) orientation of the individual layers for biaxial flexural strength disc, including the support structure (removed prior to sintering).

AM-produced parts often exhibit anisotropic mechanical strength [38]. In this investigation, each biaxial flexural test disc was printed standing on its cylinder jacket. Thus, the orientation of the individual layers is perpendicular to the direction of the load applied during the biaxial test, as can be seen in Figure 5b. In this case, the load is applied in direction of the layer boundaries; thus, this orientation typically gives the weakest possible strength values.

The hardness of the sintered silicon nitride-based ceramics was also tested with the same sample design using a test load of 10 kgf, as according to Vickers (HV$_{10}$). Ekström et al. have declared that dense SiAlON materials have a hardness HV$_{10}$ of approximately 1500 [39].

The microstructure of the sintered silicon nitride-based ceramics and their fracture surfaces were evaluated using scanning electron microscopy, SEM (Quanta 200, FEI, USA). A light microscope (Opto, Graefelfing, Germany) was used to determine the minimum wall thickness of sintered specimens.

Thermal shock resistance is an important property for high temperature applied materials. To simulate the required conditions, the printed samples were heated up to 800 °C with a dwelling time of 3 hours and a heating rate of 10 K/min. At the end of the dwelling time, the three samples (test bars with a dimension of $5 \times 5 \times 25$ mm) were quenched in water at room temperature respectively. The thermal shock resistance test samples were evaluated using light microscopy and SEM.

The thermal conductivity (λ) was calculated from the thermal diffusivity that was measured by means of laser flash analysis (XFA 500, Gammadata Instrument AB, Uppsala, Sweden). Printed silicon nitride-based discs (Ø 22 mm with a thickness of 2.5 mm) were placed in a vertical setup that had a light source on the bottom side and a detector on the top side. The samples were heated up from the bottom side with an aid of laser pulse and a detector saved the temperature during time on the top side. The thickness of the three discs and the temperatures of the plates were used to calculate the thermal conductivity.

The X-Ray Diffraction (XRD) spectra were collected using a diffractometer (X'Pert Pro, PANalytical). The surface of the sample ($10 \times 10 \times 1$ mm) was polished to achieve a flat surface. The 2θ values between 5° and 100° were measured with Cu Kα radiation ($\lambda = 0.154$ nm). The phases were evaluated using a Rietveld refinement with HighScore software.

3. Results

3.1. Suspension Characterization

The optimal level of energy needed to print the samples was determined via the cure depth of the photocurable suspension. The result of the cure depth test can be seen in Figure 6a. Due to the high level of absorption and scattering of light by the silicon nitride particles, the cure depth is significantly lower than that of other ceramic particles such as alumina or zirconia. The binder system, which is a light-sensitive organic matrix, is based on an acrylate-based monomer system and contains some additives, such as solvents and a photoinitiator [26,30]. According to Mitteramskogler et al., the curing depth should be approximately three times thicker than the printed layers to avoid the cracks during the thermal post-processing. Based on these results, all the parts generated in this study were printed using a CeraFab 7500 system (Lithoz GmbH, Vienna, Austria) with a layer thickness of 20 μm and an energy dose of 350 mJ cm^{-2}. In addition to the cure depth, the lateral over-polymerization was measured with an aid of a light microscope (see Figure 6b; E = 350 mJ/cm^2). As it is a deviation from the target dimensions, lateral over-polymerization should be avoided in order to print precise and accurate structures [40]. Due to the pixel arrangement, lateral over-polymerization does not affect tragically in that case because in that range the surface first begins to round itself.

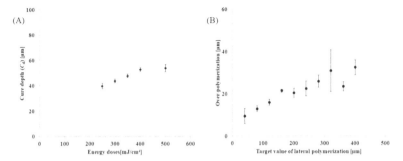

Figure 6. Cure depth (C_d) with different energy doses (**a**); lateral over-polymerization measurements (**b**) for silicon nitride-based ceramic suspension.

Figure 7 depicts the rheological behavior of the suspension at the relevant shear rates between 5 and 50 s^{-1}. Stability of the suspension was also tested by means of rheological measurements. Fresh suspension was tested (Week 0; blue dashed line) and the same suspension was also measured 8th weeks after production, both before (Week 8; red dotted dashed line) and after (Week 8*; red solid line) re-homogenization with the aid of a centrifugal mixer. Although 8-week-old suspension has a dissimilar rheological behavior to fresh suspension, centrifugal mixing helps to achieve very similar properties to the fresh suspension and can still be printed using the very same process parameters. In addition to this, all the measured viscosities have a standard deviation of less than 0.04 Pa.s.

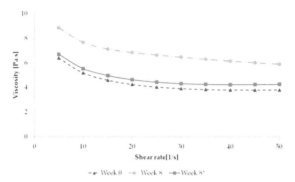

Figure 7. Viscosity of the photocurable silicon nitride-based suspension as a function of the shear rate for Week 0 and Week 8.

Finally, in terms of material efficiency, tests were performed for the three different printing cycles resulting in a variety of geometries such as test bars (5 × 5 × 25 mm) and discs (Ø 22 mm with a thickness of 2.5 mm). The amount of suspension, which was converted into shaped green parts was measured to be 89% of the LithaNit 720 suspension that was fed into the printer, which means that only approximately 13% of suspension was lost during the manufacturing of the samples (mainly due to cleaning of the green parts). This high material efficiency sets LCM distinctly apart from any other stereolithography process.

3.2. Sample Characterization

In order to gain insights into the debinding process and the effect of the debinding atmosphere, thermogravimetric analyses of the LithaNit 720 materials were carried out. Figure 8 depicts the mass changes of samples in air (blue dotted line) and in nitrogen (red solid line). The decomposition of the organic binder was completed at approximately 350 °C–400 °C in the air, whereas binder decomposition was shifted to 400 °C–450 °C in the nitrogen atmosphere, illustrating a difference in weight loss behavior due to oxidation processes. Total mass losses were 35.3% and 34.8%, respectively.

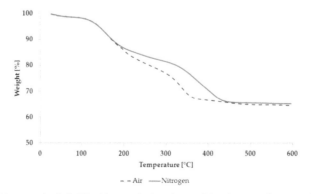

Figure 8. TGA curves for LithaNit 720 green bodies during debinding as a function of atmosphere (blue dashed line: air; red solid line: nitrogen).

The density of the sintered ceramics was determined according to the Archimedean principle by immersing in water with a measured density of 3.24 g cm^{-3}, corresponding to a relative density of 99.8 (± 0.2)% (taking into account the theoretical density reported for silicon nitride-based ceramics). These excellent densification values are comparable to conventionally produced silicon nitride-based ceramics.

Surface roughness was determined for a variety of surface orientations, as shown in Figure 9a. Measurements were performed in five different directions for three different orientated samples (namely X, Y and 45°) and the R_a values were summarized in Figure 9b. Due to the minor effect of the orientation on the building platform, the average values were noted for three different orientated samples. The average roughness R_a was 0.50 (± 0.03) μm along the layer boundaries (4) and 0.70 (± 0.04) μm perpendicular to the individual layers in the printing direction (5). The surface roughness value is a combination of the layer thickness, the angle of sample, and also the accuracy of the additive manufacturing process. Surfaces of (1) 45°, (2) top and (3) 60° show an average roughness R_a of 1.97 (± 0.11) μm, 0.76 (± 0.04) μm and 2.36 (± 0.12) μm, respectively. This can be explained by the staircase effect, which is caused by the layer-by-layer part building and also the surface angles of ceramic parts such as 45° (3) and 60° (1), resulting in a surface roughness greater than the roughness along other sample orientations [41]. However, the R_a values observed in this study are lower than values typically observed in nonlithographic AM methods [42], and are even lower than surface roughness values reported for alumina ceramics generated by LCM [30].

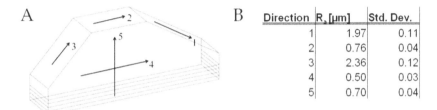

A   B

Direction	R, [μm]	Std. Dev.
1	1.97	0.11
2	0.76	0.04
3	2.36	0.12
4	0.50	0.03
5	0.70	0.04

Figure 9. (a) Directions of roughness measurement; (b) summary of surface roughness of measured samples.

Due to the elimination of edge failures by multiaxial stress, the biaxial flexural strength has advantages over uniaxial bending strength. Figure 10 shows the compressive load (N) of the tested biaxial discs, which was 1011 (± 171) N. The biaxial strength was calculated as 764 (± 137) MPa. This value can be converted into a three-point bending strength of about 950 MPa, which is the same value as for conventionally manufactured iso-pressed silicon nitride-based ceramics 945 MPa [43]. Although the printing direction was the nominally weakest orientation, the mechanical properties are very similar to conventionally iso-pressed silicon nitride-based ceramics. In addition to strength, the hardness of the biaxial discs was also measured. The resulting value of HV$_{10}$ 1500 is comparable to values reported for conventionally produced silicon nitride-based ceramics.

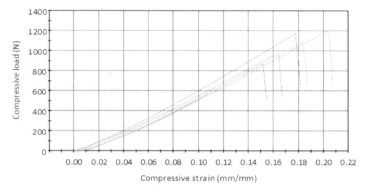

Figure 10. Biaxial flexural strength of sintered silicon nitride-based ceramics.

After determining the biaxial flexural strength and hardness of the sintered silicon nitride-based ceramic discs, fracture surfaces of the disc samples were investigated via SEM. Figure 11 depicts the fractography of samples in different magnifications. No residual layer structures introduced during the AM process were visible. Furthermore, no microcracks were observed, resulting in a smooth fracture surface. As a result of the successful consolidation process, no residual porosity was found (see Figure 11c).

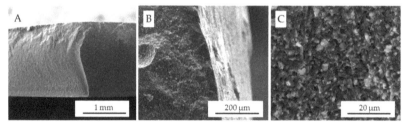

Figure 11. Scanning electron micrographs of biaxial flexural disc fracture surfaces of sintered silicon nitride-based ceramic specimens, in different magnifications with a scale bar of (**a**) 1 mm; (**b**) 200 μm; (**c**) 20 μm.

In order to inspect the minimum wall thickness, a highly macroporous gyroid-type design was printed. Figure 12 illustrates that sintered silicon nitride-based ceramics can be manufactured with a wall thickness of 265 μm without visible crack formation. The scaffold designs with a wall thickness up to 100–300 μm are commonly used for biomedical investigations [44]. For that reason, studies towards further reduction of minimum wall thickness to well below 200 μm are currently underway.

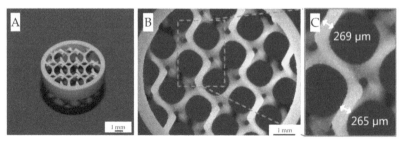

Figure 12. Printed and sintered silicon nitride-based ceramics using the LithaNit 720 suspension: (**a**) gyroids design, (**b**) and (**c**) light microscope images in different magnifications.

The thermal shock resistance of the silicon nitride-based ceramics generated by LCM was investigated by water quenching from 800 °C to room temperature. The microstructures of these test bars can be seen in Figure 13a (before water-quenching) and Figure 13b (after water-quenching). Despite the high thermal stress, no visible microcracks or other microstructural changes such as oxidation were observed.

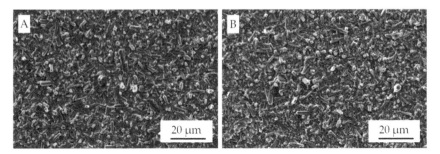

Figure 13. SEM micrographs of printed & sintered samples (**a**) before and (**b**) after thermal shock resistance test (water-quenching from 800 °C).

Thermal conductivity of conventional SiAlON grades typically varies between 20–40 Wm^{-1}K^{-1} for [3,43]. The thermal conductivity of the printed silicon nitride-based ceramic parts within this study was measured as 28.2 (± 0.4) Wm^{-1}K^{-1} using the laser flash method, which is a reasonable value for industrial applications, such as high frequency circuit substrates with ultimate thermal requirements.

The XRD analysis of the silicon nitride-based ceramics is shown in Figure 14. According to diffractogram by far the largest part of the sintered ceramic consists primarily of β-SiAlON. The sample also contains traces of a pure Y_2O_3 and pure Si_3N_4. These findings are not surprising since yttria is the most prominent sintering aid responsible for the formation of the glass phase. Due to the nature of liquid phase sintering, it is possible to find traces of pure Si_3N_4.

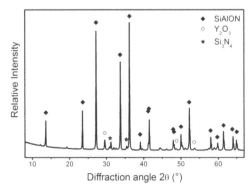

Figure 14. XRD analysis of the sintered silicon nitride-based ceramics.

4. Conclusions

In the present study, a variety of silicon nitride-based ceramic parts with complex architecture were manufactured for the first time using the lithography-based ceramic manufacturing (LCM) method, achieving material characteristics comparable to materials obtained by conventional ceramic manufacturing methods (Figure 15). After additive manufacturing using a newly developed photopolymer-containing suspension, SiAlON ceramic bodies with a relative density of 99.8% were obtained, exhibiting a biaxial strength of 764 MPa and a hardness of 1500 (HV$_{10}$). The advantages of LCM technology, such as the ability to fabricate highly precise and small complex ceramic parts, have the potential to entirely innovate the shaping of nonoxide ceramic materials in general and silicon nitride-based materials in particular. This process enables the fabrication of highly complex components (such as the ones shown in Figure 15, Figure 16), which cannot be obtained by conventional processing techniques, such as insulators, springs, impeller, microturbines and cutting tools. Additionally, due to the superb biocompatibility and good osseointegration and anti-infective (bactericidal) properties, silicon nitride makes a perfect candidate for dental (Figure 16b), orthopedic (Figure 16e) and craniomaxillofacial implants.

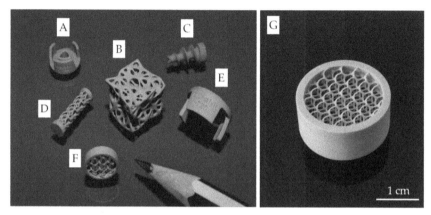

Figure 15. Printed and sintered silicon nitride-based ceramic demonstrators: insulator (**a**) and (**e**), dummy design (**b**), screw (**c**), spring (**d**), cellular designs (**f**) and (**g**).

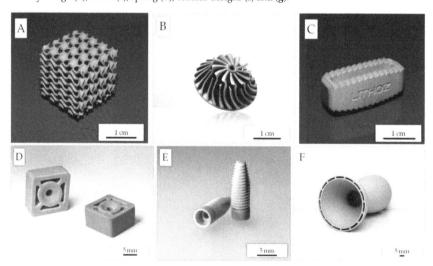

Figure 16. Printed and sintered silicon nitride-based ceramic demonstrators: (**a**) gyroids, (**b**) impeller and (**c**) spinal implant (posterior lumbar interbody fusion (PLIF) cage), (**d**) cutting tools, (**e**) dental two-piece implants with M1.6 inner thread and (**f**) de Laval nozzle.

Author Contributions: Printing, A.A.A. and T.P.; characterization, A.A.A., T.P. and T.K.; thermal post-processing, T.K.; writing—original draft preparation, A.A.A.; writing—review and editing, A.A.A., T.P., T.K. and M.S.; supervision, M.S.; project administration, M.S. All authors have read and agree to the published version of the manuscript.

Funding: This work was supported by the Austrian Research Promotion Agency (FFG) under the project number 859827 (AddiZwerk).

Acknowledgments: The authors thank Johannes Rauchenecker for the conduction of X-Ray diffraction analyses.

Conflicts of Interest: A.A.A., T.P. and M.S. are employees of the company Lithoz GmbH, supplier of the used LithaNit suspension and the CeraFab 3D printer. The authors declare no conflict of interest.

References

1. Riley, F.L. Silicon Nitride and Related Materials. *J. Am. Ceram. Soc.* **2000**, *83*, 245–265. [CrossRef]
2. Lee, M.R.; Russell, S.S.; Arden, J.W.; Pillinger, C.T. Nierite (Si3N4), a New Mineral from Ordinary and Enstatite Chondrites. *Meteoritics* **1995**, *30*, 387–398. [CrossRef]
3. Anonymous. *Brevier Technische Keramik*, 4th ed.; Fahner Verlag: Lauf, Germany, 2003.
4. Collins, J.F.; Gerby, R.W. New Refractory Uses for Silicon Nitride Reported. *J. Met.* **1955**, *7*, 612–615. [CrossRef]
5. Carle, V.; Schafer, U.; Taffner, U.; Predel, F.; Telle, R.; Petzow, G. Ceramography of High Performance Ceramics. *Prakt. Met.* **1991**, *28*, 359–377.
6. Pezzotti, G.; Oba, N.; Zhu, W.; Marin, E.; Rondinella, A.; Boschetto, F.; McEntire, B.; Yamamoto, K.; Bal, B.S. Human osteoblasts grow transitional Si/N apatite in quickly osteointegrated Si3N4 cervical insert. *Acta Biomater.* **2017**, *64*, 411–420. [CrossRef] [PubMed]
7. Webster, T.J.; Patel, A.A.; Rahaman, M.N.; Bal, B.S. Anti-infective and osteointegration properties of silicon nitride, poly (ether ether ketone), and titanium implants. *Acta Biomater.* **2012**, *8*, 4447–4454. [CrossRef]
8. Ishikawa, M.; de Mesy Bently, K.L.; McEntire, B.J.; Bal, B.S.; Schwarz, E.M.; Xie, C. Surface topography of silicon nitride affects antimicrobial and osseointegrative properties of tibial implants in a murine model. *J. Biomed. Mater. Res. A* **2017**, *105*, 3413–3421. [CrossRef]
9. Bodišová, K.; Kašiarová, M.; Domanická, M.; Hnatko, M.; Lenčéš, Z.; Varchulová Nováková, Z.; Vojtaššák, J.; Gromošová, S.; Šajgalík, P. Porous silicon nitride ceramics designed for bone substitute applications. *Ceram. Int.* **2013**, *39*, 8355–8362. [CrossRef]
10. Springer, D.B.; Parvizi, J. *Periprosthetic Joint Infection of the Hip and Knee*; Springer: New York, NY, USA, 2014.
11. Krstic, Z.; Krstic, V.D. Silicon nitride: The new engineering material of the future. *J. Mater. Sci.* **2012**, *47*, 535–552. [CrossRef]
12. Hoffmann, M.J. High-Temperature Properties of Si3N4 Ceramics. *MRS Bull.* **2013**, *20*, 28–32. [CrossRef]
13. Hampshire, S.; Pomeroy, M.J. Viscosities of Oxynitride Glass and the Effects on High Temperature Behavior of Silicon Nitride-Based Ceramics. *Key Eng. Mat.* **2005**, *287*, 259–264. [CrossRef]
14. Martin, J.W. *Concise Encyclopedia of the Structure of Materials*; Elsevier: Amsterdam, The Netherlands, 2006.
15. ASTM. *Standard Terminology for Additive Manufacturing Technologies F2792 12*; ASTM International: West Conshohocken, PA, USA, 2012.
16. Moritz, T.; Maleksaeedi, S. Additive manufacturing of ceramic components. *Addit. Manuf.* **2018**, 105–161. [CrossRef]
17. Emanuel, M.; Haggerty, J.S.; Cima, M.J.; Williams, P.A. Three-Dimensional Printing Techniques. U.S. Patent 5,204,055, 20 April 1993.
18. Subramanian, K.; Vail, N.; Barlow, J.; Marcus, H. Selective laser sintering of alumina with polymer binders. *Rapid Prototyp. J.* **1995**, *1*, 24–35. [CrossRef]
19. Bertrand, P.; Bayle, F.; Combe, C.; Goeuriot, P.; Smurov, I. Ceramic components manufacturing by selective laser sintering. *Appl. Surf. Sci.* **2007**, *254*, 989–992. [CrossRef]
20. Bartolo, P.J.; Mitchell, G. Stereo-thermal-lithography: A new principle for rapid prototyping. *Rapid Prototyp. J.* **2003**, *9*, 150–156.
21. Das, A.; Madras, G.; Dasgupta, N.; Umarji, A.M. Binder removal studies in ceramic thick shapes made by laminated object manufacturing. *J. Eur. Ceram. Soc.* **2001**, *23*, 531–534. [CrossRef]
22. Cesarano, J.; Calvert, P.D. Freeforming Objects with Low-Binder Suspension. U.S. Patent 6,027,326, 22 February 2000.
23. Grida, I.; Evans, J.R.G. Extrusion freeforming of ceramics through fine nozzles. *J. Eur. Ceram. Soc.* **2003**, *23*, 629–635. [CrossRef]

24. Zocca, A.; Colombo, P.; Gomes, C.M.; Günster, J. Additive manufacturing of ceramics: Issues, potentialities and opportunities. *J. Am. Ceram. Soc.* **2015**, *98*, 1983–2001. [CrossRef]

25. Inamura, C.; Stern, M.; Lizardo, D.; Houk, P.; Oxman, N. Additive manufacturing of transparent glass structures. *3D Print. Addit. Manuf.* **2018**, *5*, 269–284. [CrossRef]

26. Griffith, M.L.; Halloran, J.W. Freeform Fabrication of Ceramics via Stereolithography. *J. Am. Ceram. Soc.* **1996**, *79*, 2601–2608. [CrossRef]

27. Bae, C.J.; Halloran, J.W. Integrally Cored Ceramic Mold Fabricated by Ceramic Stereolithography. *Int. J. Appl. Ceram. Technol.* **2011**, *8*, 1255–1262. [CrossRef]

28. Schatt, W. *Sintervorgaenge-Grundlagen*; VDI Verlag GmbH: Düsseldorf, Germany, 1992.

29. Pfaffinger, M.; Mitteramskogler, G.; Gmeiner, R.; Stampfl, J. Thermal debinding of ceramic-filled photopolymers. *Mater. Sci. Forum* **2015**, *825*, 75–81. [CrossRef]

30. Schwentenwein, M.; Homa, J. Additive Manufacturing of Dense Alumina Ceramics. *Int. J. Appl. Ceram. Technol.* **2015**, *12*, 1–7. [CrossRef]

31. Harrer, W.; Schwentenwein, M.; Lube, T.; Danzer, R. Fractography of zirconia-specimens made using additive manufacturing (LCM) technology. *J. Eur. Ceram. Soc.* **2017**, *37*, 4331–4338. [CrossRef]

32. Pfaffinger, M.; Hartmann, M.; Schwentenwein, M.; Stampfl, J. Stabilization of tricalcium phosphate slurries against sedimentation for stereolithographic additive manufacturing and influence on the final mechanical properties. *Int. J. Appl. Ceram. Technol.* **2017**, *14*, 499–506. [CrossRef]

33. Schmidt, J.; Altun, A.A.; Schwentenwein, M.; Colombo, P. Complex mullite structures fabricated via digital light processing of a preceramic polysiloxane with active alumina fillers. *J. Eur. Ceram. Soc.* **2019**, *39*, 1336–1343. [CrossRef]

34. Zanchetta, E.; Cattaldo, M.; Franchin, G.; Schwentenwein, M.; Homa, J.; Brusatin, G.; Colombo, P. Stereolithography of SiOC Ceramic Microcomponents. *Adv. Mater.* **2016**, *28*, 370–376. [CrossRef]

35. Koopmans, R.J.; Nandyala, V.R.; Pavesi, S.; Batonneau, Y.; Beauchet, R.; Maleix, C.; Schwentenwein, M.; Altun, A.A.; Scharleman, C. Comparison of HTP catalyst performance for different internal monolith structures. *Acta Astronaut.* **2019**, *164*, 106–111. [CrossRef]

36. Chung, K.; Nenov, N.S.; Park, S.; Park, S.; Bae, C.J. Design of Optimal Organic Materials System for Ceramic Suspension-Based Additive Manufacturing. *Adv. Eng. Mater.* **2019**, *21*, 1900445. [CrossRef]

37. Philipp, H.R. Optical properties of silicon nitride. *J. Electrochem. Soc.* **1973**, *120*, 295–300. [CrossRef]

38. Felzmann, R.; Gruber, S.; Mitteramskogler, G.; Tesavibul, P.; Boccaccini, A.R.; Liska, R.; Stampfl, J. Lithography-based additive manufacturing of cellular ceramic structures. *Adv. Eng. Mater.* **2012**, *14*, 1052–1058. [CrossRef]

39. Ekström, T.; Olsson, P.O.; Holmström, M. O'-sialon ceramics prepared by hot isotactic pressing. *J. Eur. Ceram. Soc.* **1993**, *12*, 165–176. [CrossRef]

40. Mitteramskogler, G.; Gmeiner, R.; Felzmann, R.; Gruber, S.; Hofstetter, C.; Stampfl, J.; Ebert, J.; Wachter, W.; Laubersheimer, J. Light Curing Strategies for Lithography-based Additive Manufacturing of Customized Ceramics. *Addit. Manuf.* **2014**, *1*, 110–118. [CrossRef]

41. Gibson, I.; Rosen, D.W.; Stucker, B. *Additive Manufacturing Technologies Rapid Prototyping to Direct Digital Manufacturing*; Springer: New York, NY, USA, 2010.

42. Fox, J.C.; Moylan, S.P.; Lane, B.M. Effect of process parameters on the surface roughness of overhanging structures in laser powder bed fusion additive manufacturing. *Procedia CIRP* **2016**, *45*, 131–134. [CrossRef]

43. Cother, N.E. Syalon ceramics—Their development and engineering applications. *Mater. Des.* **1987**, *8*, 2–9. [CrossRef]

44. Karageorgiou, V.; Kaplan, D. Porosity of 3D biomaterial scaffolds and osteogenesis. *Biomaterial* **2005**, *26*, 5474–5491. [CrossRef]

MDPI

St. Alban-Anlage 66

4052 Basel

Switzerland

Tel. +41 61 683 77 34

Fax +41 61 302 89 18

www.mdpi.com

Applied Sciences Editorial Office

E-mail: applsci@mdpi.com

www.mdpi.com/journal/applsci

Lightning Source UK Ltd.
Milton Keynes UK
UKHW050331090722
405579UK00002B/77